SpringerBriefs in Public Health

For further volumes:
http://www.springer.com/series/10138

David Fairman · Diana Chigas
Elizabeth McClintock · Nick Drager

Negotiating Public Health in a Globalized World

Global Health Diplomacy in Action

 Springer

Prof. David Fairman
Consensus Building Institute
238 Main Street
Cambridge, MA 02142
USA
e-mail: dfairman@cbuilding.org

Prof. Diana Chigas
CDA-Collaborative Learning Projects
17 Dunster Street
Cambridge, MA 02138
USA
e-mail: diana.chigas@tufts.edu

Prof. Elizabeth McClintock
CM Partners LLC
50 Church Street
Cambridge, MA 02138
USA
e-mail: lmcclintock@cmpartners.com

Prof. Nick Drager
Professor of Practice of Public Policy and
    Global Health Diplomacy
McGill University
3430 McTavish Street
Montreal, QC X3A 1X9
Canada
e-mail: nick.drager@mcgill.ca

ISSN 2192-3698
ISBN 978-94-007-2779-3
DOI 10.1007/978-94-007-2780-9
Springer Dordrecht Heidelberg London New York

e-ISSN 2192-3701
e-ISBN 978-94-007-2780-9

Library of Congress Control Number: 2011942406

Printed on acid-free paper

Springer is part of Springer Science+Business Media (www.springer.com)

Preface

In our turbulent, interconnected and interdependent world, one person's security impacts another's, and one person's health or ill health can affect another's. Today's global economy binds the prosperity of countries together like never before, as seen in the sharp rise in food and energy prices worldwide, the decline of stock markets and the global financial crisis. These "crises of interdependence" have important public health implications. The damage is sometimes direct, as when climate change fosters the spread of communicable diseases or when rising food prices contribute to malnutrition. The damage can also be indirect, as these crises undermine the political and economic conditions necessary to promote and protect health equitably within and among societies. In a globalizing world, even local health decisions can have global impacts. At the same time, however, globalization also provides us with the opportunity to spread knowledge and resources across the world at a magnitude and with an efficacy that was impossible in the past.

Health has long been considered an issue of great "international" importance, but the recognition of health as a "global" concern is more recent. It is also profound and significant. This recognition not only requires us to think about health issues as global issues, but also alerts us to the necessity of making some health policy decisions at the global level. This is not the level at which health policy typically resides. Most health decisions have been made—and still should be made—at the local and national levels. The context of local realities has been and will remain paramount in health decision-making. To address the major health crises of today, however, and to prevent or mitigate them in the future, countries must seek collective agreement and action within *and across* their borders. As the number and frequency of decisions requiring global coordination and cooperation increase, we find ourselves in a new era of global health diplomacy.

In this new era, the most important and most commonly used tool for decision-making is negotiation. Because the international system operates on the assumptions of sovereign equality, cooperation and collaboration, rule-making, goal-setting and planning to address global health issues all require negotiation. In fact, we are currently experiencing a "new wave" of global health-related negotiations. Some of these negotiations directly address health-related issues, such as infectious disease, while others address issues such as trade and environmental policy that may have indirect but profound impacts on health policy.

In our work with health policy-makers and practitioners around the world, it has become evident to us that health decision-makers are not fully prepared for the challenges of this new era of global health diplomacy nor as skilled in the tools required to participate effectively in it as they would like. Very often, the world of international negotiation is not the world in which health decision-makers reside or are most comfortable. We believe, however, that negotiation skills are now a critical part of any health policy-makers' toolbox. Such skills are essential for diplomats, ministers, foreign policy-makers and trade negotiators. They have become equally essential for health policy makers, health practitioners and program managers within donor, governmental and nongovernmental organizations. The goal of this guide is thus to provide these actors—with particular focus on health policy-makers in developing countries—with practical information and insight into the negotiation process, so that they may be able to work toward achieving better outcomes for public health.

Acknowledgments

We would like to thank David Hermann and Kate Harvey for their tremendous work in pulling together the tools and materials for the book. We would also like to thank Anand Balachandran, Allison Berland, Kelly Sims-Gallagher, Mariya Kravkova and C. Michael Roh for their thoughtful research and analysis that resulted in the three case studies included in this guide. Special thanks goes to the WHO for its support for the research and writing of this guide, as part of its commitment to strengthening the capacities of developing country health officials to participate effectively in global health policy making processes. Finally, we would like to thank Larry Susskind and Adil Najam for their support and advice; the experience and wisdom they have offered to us has been invaluable.

Contents

Acronyms

AOSIS	Association of small island states (coalition in the climate change negotiations)
ARV	Anti retroviral
BATNA	Best alternative to a negotiated agreement
CDM	Clean development mechanism (for funding carbon emissions reductions under Kyoto protocol)
CITES	Convention on international trade in endangered species of wild fauna and flora
ETH	Ethics, trade, human rights and health law program (of the World Health Organization)
FCA	Framework convention alliance (alliance of non-governmental negotiations to enhance civil society participation in FCTC negotiations)
FCTC	Framework convention on tobacco control
IGWG	Intergovernmental working group on public health, innovation and intellectual property (in the World Health Organization)
IHR	International health regulations
IPCC	Intergovernmental panel on climate change
JFF	Joint fact-finding
MDGs	Millennium development goals
MOH	Ministry of Health
NGO	Nongovernmental organization
PhRMA	Pharmaceutical manufacturer's association (US)
SARS	Severe acute respiratory syndrome
TFI	Tobacco free initiative (of the World Health Organization)
TRIPS	Trade-related aspects of intellectual property rights
UN	United Nations
UNCTAD	United Nations conference on trade and development
USTR	United States trade representative

VDA	Virtual development academy of the United Nations development program
WHA	World Health Assembly (of the World Health Organization)
WHO	World Health Organization
WTO	World Trade Organization
ZOPA	Zone of possible agreement

Chapter 1
Introduction

Abstract Good public health generates multiple benefits for a nation's security, stability, economic well-being and relations with other countries. The public health principles of prevention, protection, accountability and equity have broad political, economic and social power. Resilient public health emerges from embedding health in *all* policies. Resilient public health supports strong national health systems, primary health care strategies and effective international and global cooperation on transnational health threats. Resilient public health is simply an integral part of good governance, whatever the political context.

Keywords Public health · Global health diplomacy · Negotiation skills · Collective action · World Health Organization (WHO) · Globalization · Negotiation · Health policy makers

Good public health is also essential for effective sustainable development. Sustainable development depends on the ability of countries, individually and collectively, to maintain and increase the stock of capital—human and social as well as physical and financial. Increasingly, human capital (people's skills, knowledge and productivity) rather than physical capital (natural resources) is the basis of comparative and competitive advantage in the developed world. In many developing countries, however, health hazards and poor sanitary conditions threaten human capital and lead to loss of life, human misery, continued poverty and underdevelopment. Poor health conditions also take a toll on social and financial capital, as social capital is increasingly consumed in addressing the unmet health needs of the poor, and as the productive forces of society that might otherwise have been employed to create financial well-being are sapped by ill health.

Globalization has intensified the public health challenges that countries face, while also constraining their ability to deal with them. Globalization has exposed countries to public health risks that were previously nonexistent or latent. It has introduced or intensified the cross-border transmission of diseases such as HIV/

D. Fairman et al., *Negotiating Public Health in a Globalized World*,
SpringerBriefs in Public Health, DOI: 10.1007/978-94-007-2780-9_1,
© The Author(s) 2012

AIDS, tuberculosis and malaria. It has increased the cross-border transmission of risk factors such as pollution, potentially unsafe foods and environmental phenomena such as climate change. It has also strengthened trade-, marketing- and travel-related cross-border influences on health behavior, especially the production and consumption of unhealthy goods and services (such as tobacco).

Globalization's impact on public health has led to the increased prominence of health on the international agenda (Drager and Sunderland 2007). Health is now competing successfully with other major issues for attention on the global stage, and indeed has risen to the top of many policy-makers' agendas. Consider the following text from the 2007 Oslo Ministerial Declaration on Global Health:

> In today's era of globalization and interdependence there is an urgent need to broaden the scope of foreign policy. Together, we face a number of pressing challenges that require concerted responses and collaborative efforts. We must encourage new ideas, seek and develop new partnerships and mechanisms, and create new paradigms of cooperation. We believe that health is one of the most important, yet still broadly neglected, long-term foreign policy issues of our time. Life and health are our most precious assets. There is a growing awareness that investment in health is fundamental to economic growth and development. It is generally acknowledged that threats to health may compromise a country's stability and security. We believe that health as a foreign policy issue needs a stronger strategic focus on the international agenda. We have therefore agreed to make impact on health a point of departure and a defining lens that each of our countries will use to examine key elements of foreign policy and development strategies, and to engage in a dialogue on how to deal with policy options from this perspective (Ministers of Foreign Affairs 2007).

This Declaration was issued and signed by the Ministers of Foreign Affairs of Brazil, France, Indonesia, Norway, Senegal, South Africa and Thailand. Together with the Doha Declaration on TRIPS and Public Health, this document represents a watershed moment for health policy-makers worldwide. Clearly, world leaders now recognize that trade needs to be managed in ways that are sensitive to health promotion, that nations must be able to deal with the health effects of global trade and, perhaps most importantly, that health is a foreign policy priority and not just an item for countries' domestic agendas. In part because of concerns such as the HIV/AIDS pandemic, Severe Acute Respiratory Syndrome (SARS), climate change and bioterrorism, health is no longer considered just "low politics"—an issue of human dignity—but rather a security, trade, economic and sustainable development issue in the realm of "high politics" (Owen 2005; Fidler 2007).

Actors in other sectors—such as trade and environment—are increasingly recognizing the importance of health to the achievement of their own development and economic goals as well. For example, former United Nations (UN) Secretary General Kofi Annan emphasized the importance of health to the attainment of the UN's Millennium Development Goals (MDGs).[1] And Bill Gates, one of the world's wealthiest individuals, and his wife Melinda created a foundation that

---

[1] In his 2001 speech calling for the establishment of the Global Fund for AIDS, Tuberculosis and Malaria, the UN Secretary General noted that "disease, like war, is not only a product of underdevelopment. It is also one of the biggest obstacles preventing our societies from developing as they should" (Annan 2001).

seeks specifically to improve health and health outcomes, with an eye toward positively impacting the economies of developing and least-developed countries. In short, the broader donor community has begun to recognize the inextricable link between health and economic development.[2]

The growing importance of health as a global policy issue is also reflected in the expanding scope and impacts of the activities of health-related multilateral institutions. In particular, the World Health Organization (WHO) has leveraged health's increasing political significance to achieve historic global health agreements—for example, the Framework Convention on Tobacco Control (FCTC, 2003) and the International Health Regulations (2005). Public health officials are playing an increasingly important role in influencing the course of events. They are shaping and managing processes of global change by devising new ways forward in implementing the MDGs; crafting global strategies on public health, innovation and intellectual property; formulating global initiatives on diet, nutrition and chronic disease prevention; and working toward global consensus on sharing influenza virus samples and the benefits that research on such samples generates.

At this inflection point in the early 21st century, the public health community has the opportunity and the responsibility to ensure that the health of populations and individuals becomes and remains a key priority of governments as they respond to the challenges of interdependence. Foreign policy and diplomacy can be "health multipliers" when solutions to international problems acknowledge and address direct and indirect threats to human health.

Translating this heightened attention on health into collective action is more problematic, however, as governments' ability to deal with cross-border public health issues is limited in several ways. First, states' capacity to influence health determinants and outcomes cannot be assured through domestic action alone. Health problems and the keys to their resolution now cut across national boundaries and often need international global solutions (Chigas et al. 2007). Second, the traditional biomedical approach to health, emphasizing disease-focused research and policy, is no longer sufficient. Gone are the days when a health crisis was the purview of a medical doctor or a health minister and his or her team. As we emphasize in an earlier article (Chigas et al. 2007: 326:): "[H]ealth problems are no longer 'only' health problems and are no longer the domain of 'only' health officials. The emerging health crises tend to be cross-sectoral crises that spill over into or are spilled over from [other] areas." Trade, environment, economic and social policies can undermine attempts to deal with health needs, and efforts to address health needs can have negative impacts in other sectors. Tourism, for example, declined significantly after China, the WHO and national governments

---

[2] The Foundation's mission is "to help all people lead healthy, productive lives. In developing countries, it focuses on improving people's health and giving them the chance to lift themselves out of hunger and extreme poverty. In the United States, it seeks to ensure that all people—especially those with the fewest resources—have access to the opportunities they need to succeed in school and life. Bill and Melinda Gates Foundation, Factsheet, http://www.gates foundation.org/about/Pages/foundation-fact-sheet.aspx (accessed April 19, 2010).

issued alerts about SARS (Ng 2003; McKercher and Chon 2004). Third, public health policy making is no longer solely within the purview of government. In order to develop and implement effective policy responses to health risks, health professionals must deal with an increasingly complex web of state and non-state actors with whom they have limited influence (Dodgson et al. 2002), but who also provide opportunities for increased resources and action on health policy. These actors are often organized trans-nationally, and have become increasingly important to agenda setting, knowledge development and dissemination, and monitoring of the health effects of non-health policies.

In short, health policy-making is now a global enterprise. There is now a complex set of goals and institutions for addressing health issues at the global level, and such issues are increasingly dealt with through foreign policy and diplomatic channels. It is unclear whether the ongoing development of global health institutions can keep pace with the spread of global threats or the inter-twining of policy areas and actors. What is clear, however, is that health negoti-ations will be more frequent, more complex and more challenging to address in a globalizing world.

1.1 Challenges for Developing Countries

Currently many countries, particularly in the developing world, are "rule-takers" and not "rule-makers" in the international system. Health policy-makers find themselves even more disadvantaged. Health policy-makers and practitioners, particularly in developing countries, are not normally at the center of international trade or development debates. They often are not included in negotiations at the national level when decision-makers are crafting their development strategies. Even within the WHO, policy-makers have had to fight to bring these issues to the attention of the more powerful member states. The WHO's Resolution WHA59.26, adopted in May 2006, for example, was a critically important step in empowering the organization to focus more actively on providing the poorest of its members with the support to negotiate effectively responses to their public health needs in an era dominated by international trade concerns:

> The Fifty-ninth World Health Assembly, having considered the report on international trade and health...URGES Member States: (1) to *promote multi-stakeholder dialogue* at the national level to consider the interplay between international trade and health; ...REQUESTS the Director-General...(2) to *respond to Members States' requests for sup-port of their efforts to build the capacity to understand* the implications of international trade and trade agreements for health and *to address relevant issues* through policies and legis-lation that take advantage of the potential opportunities, and address the potential challenges, that trade and trade agreements may have for health..." (WHO 2006a) (emphasis added).

The Resolution recognized the tremendous challenges all health policy makers face in managing the health and trade relationship and the health and foreign policy interface, as well as their need simultaneously to work internally (in their own countries) and trans-nationally in a coordinated fashion. The challenge to developing country health policy-makers is made more difficult by the fact that not all countries have equal skills and resources for participating in the complex negotiation dynamics of the new global health diplomacy.

For health policy-makers, it is not just a matter of increasing knowledge and analytic capacity, although this is important. As we have argued elsewhere, "Achieving more health-friendly negotiated outcomes is not simply a question of enhancing technical capacity to develop, monitor and evaluate health programs, increasing technical knowledge of trade rules and other areas that affect health, or enhancing research on the impacts of globalization on health. While capacity-building in these areas is essential, it is insufficient" (Chigas et al. 2007: 328). Given the new realities described above, health officials' ability to improve health outcomes is directly related to their capacity to participate effectively in negotiations and consensus-building processes in a range of policy areas and diplomatic channels. These processes take place at both the domestic and international levels and, increasingly, within forums that are not necessarily health-related but that have significant impact on health policy. In these forums, understaffed and under-resourced health ministry officials must promote and seek collective action on policy developments that are sensitive to health. The forums might include, for example, negotiations on intellectual property issues at the WTO, on sovereignty issues relating to the Convention on Biological Diversity, or on agricultural provisions within discussions of biofuels, possibly within the context of climate change policy. What happens in these negotiations has deep impacts on global health policy, but health policy-makers either find themselves absent from these forums or not fully empowered to participate and contribute fully. Health policy practitioners from developing countries find themselves doubly marginalized—not only are they least likely to participate in these discussions, they are also least likely to have the capacity to do so meaningfully.

While a variety of capacity deficits are worthy of attention, our focus in this guide is on the negotiating capacity of developing country health policy-makers. Negotiations in this new era of global health diplomacy are inherently complex. They require much preparation and demand effective and sustained management. In achieving the health outcomes they seek, health decision makers—particularly from developing countries—often face the following three types of negotiation challenges:[3]

1. **Negotiating "up"** to shape rules and actions at the global level. Ministries of health, which have traditionally focused on protecting their populations from outside health risks, must now proactively seek to influence international and global rules and actions that have spillover effects in their countries. They must understand and navigate not only global health negotiations within the framework of the WHO, but also non-health negotiations in global and regional trade, environment, foreign policy and security forums, on issues as diverse as intellectual property protection, foreign investment and trade.

---

[3] These three key elements were first articulated in Chigas et al. 2007.

2. **Negotiating "across"** to achieve national policy coherence. Health professionals must negotiate more effectively to integrate health issues and goals with their own countries' trade, development and investment policies. As the health of national populations is increasingly affected by rules and institutions in which non-health ministries and agencies take the lead, ministries of health must engage in national negotiations to shape their countries' strategies. Health professionals need to identify and influence the behavior of people in other sectors who have different worldviews, different priorities and different "cultures" and who frequently have more power and access to international rule-making forums than do health officials. They cannot simply complain about trade (Fidler 2007). They need to offer constructive solutions to make health and trade, and other sectors, work more productively together for the good of the country.
3. **Negotiating "out"** to build coalitions with diverse actors. With the growth in the number and influence of non-state actors, health professionals must negotiate and build coalitions with an increasingly broad set of actors in order to achieve health goals. The array of actors and processes is diverse and messy, ranging from local and international nongovernmental organizations (NGOs) and multinational companies to academics, scientists and professional organizations. Pharmaceutical manufacturers, health and development advocacy and service organizations, and private health management and insurance companies can all be key allies or obstacles to progress on developing country health goals. At the same time, the sheer diversity of non-state actors with an interest in health may actually make it more difficult to take action. Scientists and technical experts may align with advocates for different viewpoints and form rival epistemic communities that do not share beliefs about cause and effect or value systems that would inform them about whether and how to take action.

At all these levels, health policy-makers are being required to engage in multi-stakeholder negotiations that include the general public, ministries of trade and planning, businesses with global interests, donors and international organizations beyond the WHO. They must also exercise the skills of global health diplomacy and, perhaps most importantly, learn to more effectively shape and manage the global policy environment for health.

1.2 The Importance of Negotiation Skills

Given the global nature and breadth of the crises currently facing countries, global health diplomacy has never been more important than it is today. It is essential that health practitioners and policy-makers acquire the negotiating skills necessary to craft agreements that contribute to good public health outcomes.

How can developing countries participate effectively in negotiating such agreements? With small delegations, limited research capacity, few resources and big power differentials with many of the developed countries, the challenge is daunting. Some resources are available to developing countries—for example, information-sharing and support from NGOs such as the Third World Network and the South Centre and multilateral agencies like the WHO, among others. The WHO, in particular, through its Ethics, Trade, Human Rights and Health Law Program (ETH), supports developing countries by fostering effective global and intergovernmental action for health. The ETH provides guidance to member states and policy-makers on how to integrate ethics, human rights, social determinants and equity into policy. Through research and knowledge sharing, the ETH seeks to identify major global changes that are likely to affect public health and works with other WHO departments and partners to design strategies and possible collective action to improve public health outcomes. The ETH also provides training in health diplomacy, reinforcing the skills of policy-makers from member states in negotiation and relationship management.[4]

This last activity is a critical piece in the capacity-building agenda of the WHO and is the driving force behind this guide. The authors of this guide are motivated by the desire to better prepare health policy-makers—especially those in the developing world—for the evolving realities of global health diplomacy. As new rules, markets, actors and tools help to create a wave of new global health diplomacy, a new generation of health policy-makers must be able to facilitate, support and craft the necessary diplomacy to ensure that globalization is harnessed to deliver better health for those currently left behind in the development process. Within this context, five dimensions of negotiation are of particular importance in enhancing developing countries' leverage:

1. **Issue framing**. The pre-negotiation phase is critical for developing countries. If developing countries can get in early to frame the definition of the problem and the terms of collective debate, they can have enormous influence on the subsequent negotiation and its outcome. This is also a phase in which more "powerful" countries may not fully have formulated their views on an issue; developing countries may thus have an opportunity to influence their perspectives on the problem. Finally, this phase involves the first interaction between science and policy-making and the beginning of a process of making science "policy relevant."

2. **Managing the negotiation process**. When negotiating global public health issues, the stakes are usually high, and time is of the essence. Reaching agreement on a joint approach to solving the problem can be complicated by conflicting understandings of the facts, linkages among multiple issues and the diverse interests of the parties involved, and, of course, power imbalances among states, as well as among ministries. To deal with these challenges, health

---

[4] *See* http://www.who.int/eth/en/ (accessed April 19, 2010) for more information on WHO's ETH department.

policy-makers need effective strategies for both *joint fact-finding* (reaching a shared understanding of the facts) and developing a *mutual gains approach* to negotiation. The latter includes effective preparation on interests and alternatives to a negotiated agreement; joint value creation through brainstorming options; fairly distributing value using criteria; and creating effective processes for follow-through and implementation. These strategies can dramatically enhance the credibility and influence of a weak delegation.

3. **Coalition-building.** Given the challenges confronting any individual country or actor when involved in complex global public health negotiations, coalition-building strategies help policy-makers to more effectively deal with the power gaps between developed and developing countries and to advance their health agendas.

4. **Meeting implementation challenges.** Anticipating and securing the resources necessary to implement an agreement is critical to the success of an agreement. This step must begin during the negotiation process and continue throughout the life of the agreement.

5. **Managing institutional change**. Global public health negotiations are rarely, if ever, "one-off" events, but rather recur over time with a kaleidoscope of actors. By enhancing the capacity of individual policy-makers in the developing world to learn and apply lessons from negotiations, they can strengthen the institutional capacity of their governments and organizations to confront the health challenges of the future. With time, institutions have the ability to improve, their culture of negotiation, and the individuals within the institutions must lead that evolution.

These five steps do not constitute a "recipe" for effective negotiation on health issues of global concern, nor a comprehensive approach to global public health negotiations. A number of other important dimensions of international negotiation, including differences in culture and worldview, complicate both communication and the processes of developing mutually beneficial solutions with people from other countries and in other sectors. While recognizing and referring, where relevant, to these complicating factors, the authors believe these five dimensions are key points of leverage for health policy makers in preparing for and influencing a range of global health negotiations.

1.3 Organization of This Guide

The analytic frameworks, tools and general approaches presented in this guide are intended to serve as a broad guide to preparation for and participation in complex, multi-party, multi-sectoral international negotiations in a number of different contexts, to be adapted to the context in which any particular negotiation is taking place. Part I focuses on the fundamental negotiation leverage points, offering advice and illustrative examples to assist the reader in putting the skills and tools immediately to work. Part II presents three case studies commissioned for this book of successful achievement of developing country objectives in global negotiations that illustrate the effective application of the negotiation principles developed in Part I. These cases are not meant to be comprehensive overviews of the specific topic. They are snapshots of each negotiation, written to capture particularly salient points in the negotiation process and to underscore how effective use of the negotiation tools can advance the health goals of policy-makers. "Analyzing a Complex Multilateral Negotiation: The TRIPS Public Health Declaration" describes how *Issue Framing*, the first of the negotiation leverage points outlined in Part I, was critical to getting public health on the agenda in a forum dedicated to trade. In this case, the Africa Group also succeeded in maintaining the unity of its coalition to balance the greater negotiating power of developed countries and trade-related interests within the World Trade Organization. Case II, "Negotiating Access to HIV/AIDS Medicines: A Study of the Strategies Adopted by Brazil" provides an excellent example of 'negotiating out', describing how health officials built coalitions across a wide range of actors to achieve optimum results for AIDS patients, setting a worldwide precedent for negotiating with the pharmaceutical industry. The third case, "Keeping Your Head Above Water in Climate Change Negotiations: Lessons from Island Nations" recounts the experience of small island nations in negotiations on climate change. We chose to include this case, as the experience of developing countries in promoting their priorities in the climate change arena offers valuable lessons for global public health negotiations, especially on influencing global forums where significant power asymmetries exist. This case highlights the nature and importance of effective preparation in negotiation and illustrates how, even in situations of a perceived power imbalance, the "less powerful" party can significantly impact the results of the negotiation.

Finally, in the appendices, the reader will find several practical tools to assist in application of the elements of negotiation presented in Parts I and II, including a glossary of key negotiation vocabulary.

Part I
A Framework for Enhancing Leverage in Negotiations

Chapter 2
Issue Framing: Making Your Concerns a Global Priority

Abstract One of the challenges stakeholders in global public health negotiations face is how to focus media, public and policy-maker attention on a specific public health concern in a way that motivates action. Whether the issue is the threat posed by a new virus (e.g., HIV/AIDS, SARS, H5N1/avian flu, H1N1/swine flu), the impact of the WTO's TRIPS agreement on the access to essential medicines, or the marketing of unhealthy foods to children, defining the issue in a compelling manner is a key first step in any negotiating process.

Keywords Issue framing · Stakeholders · Negotiation · Negotiation process · Global Strategy on Public Health Innovation and Intellectual Property · Policy relevant · Winning coalition · Perceptions · Targeting stakeholders · Crafting the message · Timing · Forum choice · Law of the Sea Negotiations · Association of Small Island States (AOSIS) · Brazil · AIDS · United States Trade Representative (USTR) · PhRMA · TRIPS · World Trade Organization (WTO) · Currently Perceived Choice Tool · MDGs · International Health Regulations · SARS · Avian influenza · WHO's Global Influenza Surveillance Network · Convention on Biological Diversity · World Health Assembly · Framework Convention on Tobacco Control (FCTC) · Doha Public Health Declaration

The process of defining a situation, trend, risk or opportunity as something that decision-makers need to and can act upon is part of effective *issue framing*. "*Issue framing*" is a way of conceptualizing the issues in negotiation. It helps negotiators organize and process complex information by focusing on a particular aspect of an issue and providing a field of vision for the problem. In the domain of public health, effective issue framing is only partly a matter of technical analysis and interpretation. Scientific evidence of a health risk is usually necessary but not sufficient to motivate international action. What may be equally or more important is describing the issue in a way that is morally compelling, taps into emotion, arouses a sense of urgency and makes it clear that action can solve the problem. As

D. Fairman et al., *Negotiating Public Health in a Globalized World*,
SpringerBriefs in Public Health, DOI: 10.1007/978-94-007-2780-9_2,
© The Author(s) 2012

Jeremy Shiffman (2009, p. 608) notes, "the rise and fall of a global health issue may have less to do with how 'important' it is in any objective sense than with how supporters of the issue come to understand and portray its importance." For example, one could frame the issue of a product's risks by arguing that the product is a known carcinogen and should be regulated internationally. One could also argue that the product's health risks are borne by an under-informed public and the public health systems of developing countries, and that its benefits go primarily to large multinational corporations. The latter framing may be more compelling because it presents the issue as one of protecting the weak and less informed.

It is important to note that the framing of a public health issue is often contested. Proponents of action may state the case for action in stark terms, while those who oppose such action may emphasize uncertainties, the costs of action and the need for further study. In the recently completed negotiations on a Global Strategy on Public Health Innovation and Intellectual Property, for example, competing definitions of the issue were presented by the public health community and the pharmaceutical companies: addressing diseases of the poor vs. protection of intellectual property and drug innovation. The public health community advocated for new mechanisms to promote research and development of drugs for "diseases of the poor," such as drug-resistant tuberculosis—drugs for which there is neither a lucrative market nor any intellectual property incentive to engage in research and development. The initial position of the public health community was to question the applicability of intellectual property rules for production of drugs for diseases of the poor, arguing that drugs most needed by the poor would not otherwise be discovered, developed and delivered to them. Many pharmaceutical companies countered that any change in intellectual property rules would have drastic effects on drug innovation.

2.1 Why Issue Framing is Important

Issue framing is critically important to any negotiation process, for several reasons. First, issue framing is the first and necessary step in making information from biomedical research, epidemiology and related fields "policy relevant." Through this process, scientists and other technical experts can and must translate scientific evidence and risk and response assessments into problem statements and policy options that are not only intelligible, but relevant and compelling to the politicians, lawyers and social scientists who generally lead global negotiation processes.

Second, the way a problem is framed or defined will influence the entire course of the negotiation and shape the resulting agreement. Indeed, the initial definition or framing of a problem may be the single most crucial factor determining the likelihood and shape of the solution. It will determine what options are developed and put on the table and, to a large extent how well developing countries' interests can be met. In addition, the way an issue is framed will determine, to a large extent, which stakeholders will support and oppose action, and, consequently, whether developing countries can attract enough supporters to build a "winning coalition" for action favorable to health. For example, strategically framing a public health issue in a way that connects to the primary concerns of non-health stakeholders may help involve them in a coalition for action.

Third, in a context in which developing countries have limited ability to influence global policy making processes in which powerful developed countries (such as the United States or Western European countries) have an interest (Krasner 1985), issue framing may provide the main opportunity for low-resource stakeholders significantly to influence the negotiation process, especially when the issues are of concern to more powerful stakeholders (e.g., economic policy-makers in developed countries). If limited-resource stakeholders can get involved early to frame the definition of the problem and the terms of the collective debate, they can have enormous influence on the subsequent negotiations and their outcomes. Particularly when more powerful developed countries have not fully formulated or finalized their views on an issue, there may be an opportunity to influence their perspective. Once they have agreed on a definition of the problem and its priority relative to other negotiation issues, however, shifting their views may be very difficult.

2.2 Strategic Challenges for Global Health Stakeholders

In the issue-framing process, stakeholders seeking to mobilize others to act need to ask two basic questions:

1. "How can we persuade others that this is a problem that merits international action?" In other words, what kind of problem frame or way of defining and talking about the problem can help motivate parties to come together and negotiate a response?
2. "What kind of problem frame or definition will shape the negotiation process most favorably for *my* health priorities?"

These questions have no easy answers. Because of the nature of global public health challenges today – their cross-sector and cross-border impacts and the need to seek solutions that traverse these same borders—health advocates need to take into account three complicating factors in answering these questions.

First, many solutions will involve a complex package of resources, education, policies and programs that cut across sectoral lines. Thus, advocates will often need to persuade people *outside the health sector*, both nationally and internationally, to act. This requires getting into the "shoes" of counterparts with different values, different world views, different interests, different understandings of the problem and different perspectives and modes of argumentation associated with their professional education. For example, if one is trying to persuade states to adopt lower tariffs on essential drugs, it will likely not be sufficient to present the health benefits of the policy; one may have to address how such a tariff cut will affect government revenues. In this sense, it is useful to consider cross-sectoral negotiations as cross-cultural ones, with the challenges of communication, mutual understanding and bringing together different values and perspectives that these present.

Second, public health officials will likely need to overcome the common perception that incorporating health concerns entails restrictions and negative impacts on other sectors. In other words, policy-makers in other sectors may see public health regulation as an impediment to development and a limitation on free trade and business flexibility. Incorporating public health concerns is thus often seen as a win-lose proposition.

Third, public health policies may not have immediate, direct and visible benefits for other sectors. The economic benefits of public health are longer term and more indirect, even if very real. Thus, persuading others of the benefits improvements in public health will bring to their sector, whether economy, development, reduction of poverty, trade or governance, is challenging, as they may see themselves as giving up something now in return for an uncertain benefit in the future.

Health policy stakeholders' critical first hurdle is thus to reframe health problems in a way that motivates other sectors to be concerned about them. And, strategies that have been effective *within* the health sector may not be as effective in a cross-sectoral or global context. Presenting evidence of the impact of trade (or other policies) on health will be necessary, but not sufficient. In fact, heightened public concern about an issue may be more important than scientific consensus in determining whether policy-makers take an issue seriously. For health policy officials and other stakeholders who are accustomed to medical or technical responses to public health problems, this represents an enormous shift in thinking and action.

2.3 Framing Strategies

Despite the challenges, health stakeholders do have significant opportunities to influence the identification and framing of issues, and they have experienced success in the health field and other fields. Four strategies have proven to be particularly effective:

(1) Targeting the right stakeholders for action;
(2) Crafting the message for maximum influence;
(3) Timing the initiative to build momentum;
(4) Seeking a favorable forum for negotiation.

2.3.1 Targeting the Right Stakeholders for Action

The first step in framing an issue for potential international or global negotiation is to determine the target audience. One way to do this is to determine who will need to act, what action they will need to take in order to address the issue, and, if you do not have direct access to or influence on these key decision makers, it is important to determine how to influence them indirectly.

For example, in 2005, a coalition campaigning for aid to Africa determined that the G-8 leadership was their ultimate target for action because these leaders collectively could make a commitment to increase aid to Africa substantially. Having made that assessment, however, the coalition partners recognized that they had no ability to influence G-8 leaders through direct communication at the G-8 Summit. They determined that their best avenue of approach was through the citizens of G-8 countries, who could communicate their views on the need for aid to Africa through letters, e-mails and participation in public events and thereby increase pressure on the G-8 to take action. Their next step was to organize a series of rock concerts in G-8 countries, advocating for aid and giving citizens clear opportunities and instructions for contacting their leaders. Millions of citizens did so, contributing to the G-8 countries' decision to double aid to Africa by 2010.[1]

2.3.2 Crafting the Message for Maximum Influence

Once you have identified the actors you want to influence, and the "pathway" to reach them, the next step is to frame the message in a way that has the best chance of influencing them. Certain kinds of problem framing are more persuasive than others. Research on communicating global health messages to publics and policy-makers in developed countries indicates that the most effective message platform tends to:

1. Describe the problem using credible facts the audience can relate to, and, to the extent possible, describe how the problem affects the target actors, their constituencies or issues they care about.
2. Describe a viable solution to the problem with hopeful, positive, simple language and tangible examples of how the solution might work (or has worked).
3. Advocate action steps to help solve the problem, i.e. calls on target actors to do something specific.
4. Make a moral appeal for action. Messages need to connect emotionally with the target audience to inspire them to take action.
5. Make a concrete call to action (Global Health Council 2005).

---

[1] G8 Hokkaido Toyako Summit Leaders Declaration, 8 July 2008, available at http://www.mofa.go.jp/policy/economy/summit/2008/doc/doc080714__en.html.

Box 2.1 "[G]lobal polio eradication has been positioned as a humanitarian crusade to rid the world of a scourge that has afflicted children for millennia. Many older advocates from industrialized nations may view this positioning as both credible, accepting the idea that polio is truly a problem the world can be rid of, and salient, remembering a time when polio caused havoc each year in their own countries" (Shiffman 2009, p. 609).

The importance of effective message crafting is illustrated by two contrasting examples: the failure of developing countries' to achieve equitable sharing of ocean resources (in the Law of the Sea negotiations) and the success of small island states in negotiating the climate change-related Kyoto Protocol. In the Law of the Sea negotiations, developing countries insisted that seabed resources such as manganese nodules be shared according to the principles of the New International Economic Order (NIEO). The moral claim on which the NIEO was based—that developed countries owed developing countries a large debt to compensate for their exploitation during the era of colonialism – was not compelling to developed countries. In fact, this argument strengthened U.S. and other Western countries' opposition to developing countries' interests and agenda. Industry representatives with an interest in seabed mining argued that the declaration that the seabed was the "common heritage of mankind" was "collectivist," and that seabed production controls were "OPEC-like cartelization," and that mandatory technology transfer damaged intellectual property rights. They argued persuasively to the U.S. government that what was at stake in the negotiation was not only rights to seabed mining and their financial interests, but a precedent with respect to global governance (Sebenius 1991, p. 128).

In the negotiations on the Kyoto Protocol, by contrast, the Association of Small Island States (AOSIS) became an influential driver of action. AOSIS framed the problem of climate change in a way that connected inaction to very concrete, disastrous consequences (such as the complete disappearance of small island states) and created a moral dilemma for the large industrial countries. They also presented the threat of global climate change not merely as a threat to Pacific and Caribbean islands, but to the physical (and economic) integrity of the East Coast of the United States, highlighting a potential loss of immediate concern to the U.S. This framed the problem in terms of U.S. interests, not just those of small island states, and helped to bring the U.S. to the table (see Chap. 8).

Message crafting is important not only for creating a compelling argument, but also for generating persuasive options for negotiation and building alliances to achieve public health outcomes. Brazil's success in achieving lower prices for pharmaceuticals for AIDS sufferers was due in part to the Government's ability to understand the concerns of key decision makers in the United States and frame the decision in a way that made it difficult to refuse to take action. At issue was a provision in Brazil's patent law requiring local production as a condition for foreign patent holders to receive protection in the country. The United States—led by the Pharmaceutical Manufacturer's Association (PhRMA) and the U.S. Trade Representative (USTR)—viewed this as inimical to free trade and a violation of the TRIPS agreement. USTR had on previous occasions exerted significant pressure on Brazil (through the threat of trade sanctions) to change its patent laws to favor stronger protection—with success. In this instance, however, that pressure did not work. Rather than accept the US characterization of the issue at stake as the legality of its patent law and the protection of intellectual property rights, Brazil chose to frame the issue for negotiation as one of "access to essential medicines." This approach enabled Brazil to overcome the opposition of the United States to reducing prices of AIDS drugs. The President of Brazil was able to attract diverse allies within the country (including ministries, NGOs and industry groups), in the United States (including NGOs and media), and internationally (within multilateral forums and other countries). NGOs in both Brazil and the US took up Brazil's causes and worked hard to disseminate information to key decision makers and the public.[2] Articles appeared in newspapers or on the internet characterizing the US position as unethical. Gradually, public opinion began to shift towards the Brazilian side (See Chap. 7, p. 209).

Brazil's framing of the issue in terms of public health also effectively linked its situation to that of Africa, where concern about the spread of AIDS had increased substantially—so much so that in May, 2000, President Clinton issued an Executive Order declaring that the US government would not impose trade sanctions against African governments that violated American patent law in order to provide AIDS drugs at lower prices (Lewis 2000).[3] Brazil made this link explicit by underlining its negotiation with the U.S. as a precedent-setting model for Africa. "Brazil has raised this banner because it is a cause that has to do with the very survival of some countries, especially the poor ones of Africa," President Cardoso of Brazil had said in an interview with the New York Times (Petersen and Rohter 2001). This effectively made it more difficult for USTR and PhRMA to exert pressure on Brazil without significant public and international outcry.

Finally, by expanding the issue beyond a pure "price negotiation," Brazil and its allies enhanced their negotiation power by creating possibilities for new options that allowed the United States to back down from its initial position. The reframing enabled the United States to find a way to take the decision Brazil wanted without being seen as compromising on its biggest concerns: protection of intellectual property and avoidance of a bad precedent for future

---

[2] *See, e.g.,* Oxfam 2001 and Medecins Sans Frontieres 2001.

[3] The Order prohibited the US from taking action pursuant to US trade laws against any sub-Saharan African country that promoted access to HIV/AIDS drugs in a way that provided adequate protection of intellectual property in accordance with TRIPS provisions. Executive Order 13155 of May 10, 2000, "Access to HIV/AIDS Pharmaceuticals and Medical Technologies." *Federal Register.* Vol. 65, No. 93 (May 12, 2000).

negotiations. "[T]he [May 2000 executive] order strikes a proper balance between the need to enable sub-Saharan governments to increase access to HIV/AIDS pharmaceuticals and medical technologies and the need to ensure that intellectual property is protected," President Clinton commented at the time (Clinton 2000). It still required countries to provide "adequate and effective intellectual property protection", but accepted the WTO's standard on patents, rather than applying the US's more stringent rules and included some important caveats preserving the US's rights to take action (Executive Order 13155, Sect. 1(a)). The Executive Order applied only to African countries at the time, but its effects spilled over immediately to the US-Brazil dispute. Two days later, five major pharmaceutical companies agreed to negotiate voluntary price cuts in Brazil, and in June, 2001, the United States withdrew its complaint with the WTO concerning the Brazilian law.[4]

The *Currently Perceived Choice* tool (*see* Appendix 2) can assist in the process of framing issues and messages in ways that facilitate action on public health priorities—particularly in situations in which counterparts (whether other Ministries in the country or other countries in the context of an international negotiation) are refusing to take action favorable to public health. Consider the example of a situation in which the Ministry of Health of Country X, in cooperation with the WHO, would like action to be taken in neighboring Country Y to restart a vaccination campaign that was halted by Country Y. Country Y has refused access to WHO and NGOs to perform the vaccinations. Country X is concerned that without the vaccinations, the spread of disease across its border will be rapid and devastating. After some analysis, the Minister of Health of Country X and WHO determined that local authorities (and not the national Ministry of Health) should be targeted, as they initiated the prohibition of vaccines. There is overwhelming evidence of the effectiveness and safety of the vaccines—enough, they believe, to convince even the strongest skeptic. But it has not been sufficient to motivate effective action by the local authorities, even under pressure from the national Ministry of Health. The following illustrates one way in which the local official might be thinking about his choice:

---

[4] Brazil restricted the possibility of compulsory licensing to cases of a national health emergency and agreed to notify the US government in advance if it found it necessary to issue a compulsory license (Wogart et al. 2008).

Box 2.2
Currently perceived choice
Decision Maker: Local Authority in Country Y
Decision he/she believes she is being asked to take[5]: "Shall I now bow to pressure from the West and the Central Govt. to re-start a potentially dangerous vaccination campaign?

If I say "YES" (Consequences of saying "yes")	If I say "NO" (Consequences of saying "no")
- Would be seen as capitulating to the West and the Central Government	+ Would be seen as standing up to the West and the Central Government
- Will be accused by religious leaders (and my supporters) of participating in a Western plot to make our children infertile	+ Will be hailed by religious leaders as protecting the lives of our children and of future generations
- Would confirm that children were paralyzed due to ignorance and unfounded rumors, as stated in the press	+ Will be seen as standing up for fairness and promoting further transparency by the Central Government
- Would be seen as incompetent: would accept that our initial test was erroneous	+ Will be seen as supporting my personnel in our State, and not admitting a mistake
- Would admit mistake and would discredit our doctor who conducted the tests	+ I may be able to play a bigger role in next elections with support that my stand will garner
- Would be seen as not protecting the people; Central Government would get the credit from the people, other countries in the region and the global community	+ I can always agree later

[5] Reminder: The question that is being asked *may not* be the question that is being *heard* by the decision-maker. See Appendix 2 for further explanation of the Tool

In reviewing the Currently Perceived Choice, the Ministry of Health officials from Country Y and the WHO may gain insight into why the evidence they presented regarding the spread of disease and the safety of the vaccines failed to convince the local authorities. Other relevant factors affecting the local authorities' decision regarding vaccination include:

(1) The influence of religious leaders in the province. The Ministry and WHO might ask whether the religious leaders might be useful stakeholders to target, given their influence over the decision.
(2) The importance to local officials and stakeholders of participation in the process of devising and implementing solutions and of receiving credit for successes. It is important that the process not be seen as a Central Government process, but that it reinforce local credibility, authority and visibility.
(3) The importance to local leaders of being perceived as competent (or at least not incompetent), especially vis-à-vis the Central Government and the West.

With a better understanding of the factors affecting target stakeholders' perceptions of the problem, it becomes clear that having medical evidence to support the action the WHO and Ministry of Health favor will likely not be enough to generate decision makers' support, and, in this case, might even trigger negative reactions. The question might need to be reframed to target the local officials (and not the Ministry of Health), and to take account of the other factors influencing the target stakeholders' decision.

2.3.3 Timing the Initiative to Build Momentum

In addition to the knowledge of *whom* to target and *how* to craft the message, a third aspect of framing is critical: timing. Is the time right to move forward with a campaign or proposal for action? To make this assessment, health stakeholders need to answer three questions:

- Is there an opportunity to link the issue to existing global priorities and resources?
- Is there a dramatic event or situation that provides a focus for public attention and a rallying point for action?
- Is there an appropriate forum available, with a mandate that could be interpreted or stretched to include the issue?

Often, a core group of stakeholders will recognize the need to address a specific public health concern, while policy-makers and the wider public do not yet believe the issue has achieved sufficient salience to motivate action. In these situations, convergence with existing global priorities and resources may allow for linkages that lead to a tipping point. For example, climate change was initially viewed by some as predominantly an environmental problem rather than a development problem. Yet the impacts of climate change can directly affect the efficient investment of resources and the achievement of development objectives. At the same time, how development occurs has an impact on climate change and the vulnerability of societies (OECD 2005). Developing countries succeeded in reframing the issue from one of protection of the environment to "sustainable development." As the framing of the climate change problem shifts to include linkages between environment and development, the specific concerns of developing countries are gaining prominence on the climate change agenda (Chigas et al. 2007).

Similar opportunities for linkage that increase the salience of public health have arisen with regard to the UN's Millennium Development Goals. The MDGs, which were adopted as a UN General Assembly Resolution by the heads of state at the 2000 Millennium Summit, include commitments directly related to public health issues: to reduce child mortality, to improve maternal health and to halt and begin to reverse the spread of HIV/AIDS, malaria and other diseases by 2015.[5] The high global visibility of the MDGs and their prominent entrance into the development

---

[5] See www.un.org/millenniumgoals/.

debate have provided an opportunity for advocates of these public health causes to frame their concerns in ways that directly connect to the global agenda of development and poverty eradication. By linking their priorities to concrete action pledges from the MDGs public health advocates and officials can raise their prominence and argue more persuasively for action on public health.

A single dramatic new development may also create a window of opportunity to draw public attention to a public health issue and move toward action. The outbreak of the Severe Acute Respiratory Syndrome (SARS) epidemic, a human respiratory disease, in southern China in late 2002 and its subsequent rapid spread throughout the world illustrated the wide-ranging impact of a new disease in a closely inter-connected and highly mobile world. The outbreak of SARS highlighted the importance of a worldwide surveillance and response capacity to address emerging microbial threats through timely reporting, rapid communication and evidence-based action. Shortly after SARS was recognized as a threat to human health, the WHO took swift and sweeping measures. These measures included issuing global alerts that were amplified by the media and brought greater vigilance and more rapid detection and isolation of cases; direct technical support to assist in epidemiological investigations and containment operations; and the establishment of research net-works to enhance knowledge about the disease (WHO 2003). This new and emerging communicable disease threat was also a catalyst for renewing interest in completing longstanding negotiations on the revision of International Health Reg-ulations (WHO 2003b)[6] and changing the scope of these regulations from just three diseases – smallpox, plague and yellow fever—to include all "public health emer-gencies of international concern" (WHO 2005).

Similarly, the outbreak of avian flu in 2006 and a crisis that emerged from efforts to develop an international response to it drove global health stakeholders to reframe a previously deadlocked issue—whether viruses should be treated as sovereign resources or as shared international concerns—in terms of access to vaccines. During previous negotiations on the Convention on Biological Diversity, disagreements about sovereignty and ownership of biological materials had

---

[6] The International Health Regulations aim to prevent, protect against, control and provide a public health response to the international spread of disease in ways that are commensurate with and restricted to public health risks, and which avoid unnecessary interference with international traffic and trade (IHR 2005, Article 2). They are an international legal instrument that governs the roles of WHO and its member countries in identifying and responding to and sharing information about public health emergencies of international concern. The IHR build on and expand a series of regulations, the International Sanitary Regulations, adopted in 1951. In 1969, they were revised and adopted as the International Health Regulations, regulating three diseases: cholera, plague and yellow fever. The 2005 regulations include smallpox, polio, SARS and new strains of human influenza that member states must immediately report to the WHO and provide specific procedures and timelines for announcing and responding to public health events of international concern.

prevented agreement on virus sharing.[7] Indonesia, the country most affected by the H5N1 (avian flu) virus, had announced in 2006 that it would no longer share samples of the virus with WHO's Global Influenza Surveillance Network. The loss of access to H5N1 virus samples posed serious risks to global health security, because samples of the virus are essential for development of flu vaccines. In response, national governments renewed international negotiations through an *ad hoc* meeting in Jakarta.

In the negotiation process, government negotiators reframed the issue of virus sharing, shifting from a focus on sovereignty and ownership (a framing which had led to deadlock) to one of the "responsible practices for sharing avian influenza viruses and resulting benefits" (WHO 2007a).[8] This reframing allowed for recognition of a country's sovereignty over the viruses while creating an obligation to share them. That obligation was formally affirmed in a resolution of the World Health Assembly (World Health Assembly 2007).

2.3.4 Seeking a Favorable Forum for Negotiation

As the avian flu virus sharing example shows, effective negotiators may seek not only to reframe an issue, but also to shift the forum in which the issue is discussed. Making the choice of a forum part of negotiation strategy is especially useful when one forum seems more likely than another to resolve the issue in a way that meets the negotiator's interests.

In the case of the avian flu virus, the Convention on Biological Diversity had deadlocked on the question of sovereignty over biological resources. Faced with a crisis in 2006, international health negotiators pushed for the issue to be reopened in an *ad hoc* forum focused specifically on sharing the avian flu virus, and then sought ratification of their proposals not in the Convention on Biological Diversity, but rather in the World Health Assembly. For these negotiators, safeguarding public health was much more important than protecting national property rights. Therefore, they convened a forum far more focused on public health concerns than on biological property rights, and sought formal government ratification through

---

[7] The Convention on Biological Diversity (5 June 1992) defines biological resources to include "genetic resources, organisms or parts thereof, populations, or any other biotic component of ecosystems with actual or potential use or value for humanity" (article 2) and states that "the authority to determine access to genetic resources rests with the national governments and is subject to national legislation" (article 15.1). Genetic resources are defined to include "any material of plant, animal, microbial or other origin containing functional units of heredity" (article 2). *See* Fidler 2008 for an analysis of the international legal dimensions of virus sharing.

[8] In announcing Indonesia's resumption of virus sharing, Dr David Heymann, WHO's Assistant Director-General for Communicable Diseases, commented: "We have struck a balance between the need to continue the sharing of influenza viruses for risk assessment and for vaccine development, and the need to help ensure that developing countries benefit from sharing without compromising global public health security." (WHO 2007b).

the World Health Assembly, where they could be confident that public health concerns would outweigh sovereignty concerns.

Typical international forums for health authorities to negotiate rules and regulations on health issues include the WHO's World Health Assembly (WHA) and its related processes of intergovernmental meetings,[9] the governing boards of the Joint UN Programme on HIV/AIDS, the Global Fund and Global Alliance on Vaccines and Immunization, and increasingly the UN General Assembly.[10]

Ad hoc forums are gaining in importance, especially in public health. They provide an opportunity to focus attention, resources and policy-making *directly* on public health issues, rather than on public health priorities in forums in which they are a secondary concern. The 2003 Framework Convention on Tobacco Control (FCTC) is an example of an *ad hoc* negotiation process that produced the first framework treaty adopted under the auspices of the WHO. The FCTC is designed to strengthen international and national cooperation to reduce the growth and spread of the global tobacco epidemic, which disproportionately affects developing countries. It was negotiated under WHO authority and is modeled on the framework convention-protocol approach successfully utilized in international environmental law.[11] In preparation for the negotiations, the World Health Assembly established an Ad Hoc Inter-Agency Task Force on Tobacco Control under WHO leadership. Though part of WHO's mandate was to improve coordination and cooperation across UN agencies, the choice of WHO as the convener of the negotiation process also served a strategic purpose: to replace the UN Conference on Trade and Development as the UN convener. In the existing UNCTAD forum,

---

[9] *See*, for example, the Intergovernmental Working Group on Public Health, Innovation and Intellectual Property, http://apps.who.int/gb/phi/; the Pandemic Influenza Preparedness (PIP) process, http://apps.who.int/gb/pip/; the Conference of the Parties to the WHO Framework Convention on Tobacco Control (FCTC) and the Intergovernmental Negotiating Body on a Protocol on Illicit Trade in Tobacco Products, http://apps.who.int/gb/fctc/; and the WHO Global Code of Practice on the International Recruitment of Health Personnel, http://apps.who.int/gb/ebwha/pdf_files/WHA63/A63_R16-en.pdf.

[10] *See*, e.g., UN General Assembly Resolution 64/265 (13 May 2010) on the prevention and control of non-communicable diseases; UN General Assembly Resolution 63/33 (26 November 2008) (requesting the UN Secretary General to prepare "in close collaboration with the Director-General of the World Health Organization, and in consultation with Member States, to submit to the General Assembly at its sixty-fourth session, in 2009, a comprehensive report, with recommendations, on challenges, activities and initiatives related to foreign policy and global health, taking into account the outcome of the annual ministerial review to be held by the Economic and Social Council in 2009"); UN General Assembly Resolutions 63/234 (22 December 2008), 61/228 (22 December 2006), and 55/284 (7 September 2001) on "2001–2010: Decade to Roll Back Malaria in Developing Countries, especially in Africa"; Declaration of Commitment on HIV/AIDS, UN General Assembly Resolution A/RES/S-26/2 (27 June 2001).

[11] The "framework convention" approach has been used widely in climate change negotiations. It involves negotiating a general agreement that acknowledges the existence of the problem with principles for a solution, including perhaps targets for action, followed by negotiation of specific protocols with details of how the principles will be put into practice. This step-by-step approach was in part a reaction to the years of negotiation of a detailed and comprehensive Law of the Sea treaty that was ultimately rejected by the United States (Sebenius 1991, p. 14).

the tobacco industry had substantial influence on agenda setting and decision making, and had blocked meaningful action to limit trade in tobacco (Collin et al. 2002).[12] Once underway, the FCTC process gained significant political momentum and turned into a worldwide public health movement (Roemer et al. 2005).

Health issues are also increasingly negotiated in non-health *ad hoc* forums. The negotiation of the 2000 Cartagena Protocol to the Convention on Biological Diversity illustrates the increasing significance of such nontraditional forums. In order to adequately address the potential risks posed by cross-border trade and accidental releases of "living modified organisms," negotiators had to consider environmental issues as well as matters of health, food safety, trade, property rights and socioeconomic development. Even though the core issue of biosafety made environment ministers the driving force behind the negotiations, subsequent ratification and implementation demanded coordination with ministries of health, science and technology, agriculture, and trade. Those players could thus move their specific concerns onto the agenda (Martinez 2001).

Public health concerns have also featured prominently in trade negotiations. In the Doha Development Agenda, the WTO-led trade negotiations that began in November 2001, for example, many developing countries viewed an agreement on Trade-related aspects of intellectual property (TRIPS) and public health as an essential element of the trade agenda. The resulting Public Health Declaration reflected the success of developing countries in ensuring that TRIPS would be interpreted in a way that supported their public health goals.

When health negotiators are considering how best to frame an issue for action, they should also consider what forums are available for international negotiations and decisions on the issue. As the avian flu and tobacco control examples illustrate, the choice of forum may support or undermine the potential for agreements that advance the negotiator's interests. Negotiators should keep at least three criteria in mind as they consider where to "bring" an issue for international action:

1. *Likelihood that the forum will produce an agreement or decision that meets our interests.* Is this a forum where representatives are likely to share our view of the issue and to agree with our preferred course of action? If the negotiator's goal is to advance a public health interest over an economic interest, then the WHA is likely to be a more favorable forum than the WTO. However, the likelihood of support alone is not enough to drive the negotiator's decision.

---

[12] The FCTC was negotiated in six sessions of the Intergovernmental Negotiating Body, with "intersessional" consultations and individual and group consultations by the chair with various delegations, and was approved by the 56th World Health Assembly on 21 May 2003. A Conference of the Parties to follow up on the FCTC held its first meeting in 2006. It has since established working groups on different articles, as well as an Intergovernmental Negotiating Body on a Protocol on Illicit Trade in Tobacco Products.

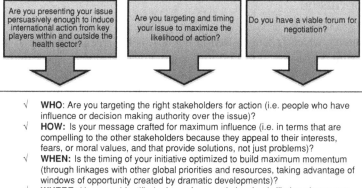

Fig. 2.1 Issue framing

2. *Likelihood that agreements/decisions in this forum will have real impact.* How directly and how strongly will agreements and decisions reached in this forum affect the outcome we are seeking? Are agreements reached here binding on actors whose behavior we are seeking to change? Are commitments made here likely to result in resources being mobilized to support action on the issue? Though a negotiator may believe that the WHA is more likely to favor his/her interests than the WTO, the issue may be one on which the WHA actually has little or no power to bind key actors, and the WTO, by contrast, does have such power. Public health advocates decided to advocate for access to pharmaceuticals through the Doha TRIPS negotiations in large part because the WTO has far more authority and resources to enforce an agreement on access to pharmaceuticals than the WHA.

3. *Our ability to participate in negotiations under the auspices of this forum:* Is this a forum where our agency/coalition can be directly represented, or will we need to work through others? If we cannot be present ourselves, do we have allies among the representatives who can effectively negotiate on our behalf? If not, do we have a way to build those alliances in a timely fashion? Particularly in the case of non-health forums, such as the WTO or the Convention on Biological Diversity, health negotiators must determine whether they can "get to the table" themselves, or whether they will need to work through government delegations that are led by trade, environment or development agencies whose primary concerns are different. Building relationships and alliances with counterparts in non-health agencies is especially important when health advocates must count on those counterparts to represent their interests, and can have only indirect participation in the negotiation process.

2.4 Conclusion

Issue framing, which is summarized in Fig. 2.1, is the crucial first step in making public health concerns a global priority and motivating action, especially because the mere existence of scientific evidence of a public health concern is usually not enough to drive policy-makers to act. By framing the problem and the terms of the broader debate, less-influential stakeholders can have a major impact on the terms of the ensuing negotiations and resulting agreement. As they seek to craft a compelling framing of the issue, negotiators also need to target stakeholders who are empowered to act, link to other high priority issues on the international agenda, and move their issue to a forum that is favorable to their interests and to action.

Chapter 3
Managing the Negotiation Process

Abstract In global public health negotiations, the stakes are usually high and often time is of the essence. The outbreak of the SARS epidemic in late 2002, for example, illustrated how rapidly crises can occur and how immediate action may be required. Negotiations on immediate and short-term issues such as SARS, and even on long-term policies not triggered by a crisis, can be made all the more complex by diverse interests, conflicting understandings of underlying facts and linkages among the multitude of issues. Specific obstacles to joint problem-solving may include disagreement on the existence, certainty or severity of the problem; on the best way to tackle the problem or the likelihood of success; or on who bears responsibility to act, who will pay costs and who will manage the response. In the health sector, national leaders in key countries may be reluctant to acknowledge the urgent need to address the spread of a disease, either because they question the facts or because they fear that taking action will have negative impacts on their international image and/or domestic political support.

Keywords Negotiation · Negotiation process · Joint fact-finding · Interests · Interest-based negotiation · Stakeholder process · HIV/AIDS · Intergovernmental Working Group on Public Health, Innovation and Intellectual Property (IGWG) · Intergovernmental Panel on Climate Change (IPCC) · World Health Assembly (WHA) · Commission on Intellectual Property Rights, Innovation and Public Health · Mutual gains approach · Preparation · Value creation · Value distribution · Follow through · Implementation · Best Alternative to Negotiated Agreement (BATNA) · Alternative · Alliance of Small Island States (AOSIS) · Doha Declaration on TRIPS and Public Health · World Trade Organization (WTO) · Positions · Indonesia · Avian flu virus · Options · Brazil · Brainstorming · FCTC · Pandemic influenza preparedness negotiations · Criteria · Framework convention on climate change · Contingent agreements · Issue mapping · Issue trade-offs · Single text approach · Modes of decision-making · Building trust · Culture

D. Fairman et al., *Negotiating Public Health in a Globalized World*,
SpringerBriefs in Public Health, DOI: 10.1007/978-94-007-2780-9_3,
© The Author(s) 2012

To deal with these obstacles, health-related negotiations must include effective strategies for reaching a shared understanding of the facts, creating options that meet the primary interests of the key stakeholders, and "packaging" options and trade-offs into agreements that stakeholders see as fair. This chapter presents an approach to negotiation that offers an alternative to "hard bargaining"—a way to overcome obstacles to agreement by focusing on producing gains for all key stakeholders.

3.1 Establishing a Shared Understanding of the Facts: Joint Fact-Finding

Negotiations on international public health issues often take place in an atmosphere of urgency and crisis. When the SARS epidemic caught the world's attention in March 2003, efforts to contain the virus were undertaken immediately and took place in a climate of continuous and intense media reporting that magnified the urgency of the threat ("Q&A: SARS" 2004).

In critical situations such as these, it is important to define and resolve technical and scientific questions to the fullest extent possible at the outset of the decision-making process, in order to avoid time loss and suboptimal outcomes. However, efforts to establish the facts and define technically feasible options for joint global action are often confounded by the problem of "dueling experts." The dueling experts problem arises when stakeholders who disagree on the basic facts of the issue and/or the effectiveness of a particular response bring forward experts in support of their respective views. This problem undermines negotiations in several ways:

- It often results in the introduction of one-sided and incompatible scientific evidence, which is seen as authoritative by its supporters and spurious by those on the other side. Experts may feel the need to defend their work and criticize the assumptions, methods and findings of their counterparts. Such polarized expert debates make it more difficult for stakeholders to come to a common reading of the facts. Experts on the effects of patents on innovation of new products and of access to these products often face conflicts of interest that undermine their claims of neutrality and objectivity in presenting scientific evidence on these issues. Experts' close ties to industry, NGOs or governments may affect others' perspectives of those experts' opinions and have a significant impact on their ability to seriously engage in a joint systematic review and discussion of the data. This can happen at both the national and international levels. In May 2003, for example, a panel convened by the U.S. National Institutes of Health recommended the broader use of hypertension drugs at lower blood pressures, but nine of the eleven authors of the guidelines had ties to drug companies (Wilson 2005).

- Less wealthy countries may not have equal access to experts. These countries may thus be at a significant disadvantage compared to rich countries, which can magnify their voices, and therefore enhance their negotiating power, through experts. And such disparities might make it easier for industrial lobbyists to influence a negotiation, despite their conflicts of interests.

The dueling experts dilemma prolongs the process of finding consensus and frustrates stakeholders who are not technical experts. This dilemma often results in suboptimal solutions based on political compromise within the range of arguments presented by the dueling experts. As an example, some believe that conflicting scientific evidence and the lack of a joint establishment of technical and scientific facts resulted in the watering down of findings presented in reports by the UN Intergovernmental Panel on Climate Change on the likely damaging impacts of climate change; in that case, language calling for cuts in greenhouse gases was eliminated at the insistence of diplomats whose pursuit of specific political agendas was facilitated by the presentation (Eilperin 2007). Conversely, email comments by climate scientists suggesting that they had avoided publishing data that did not support their global warming hypotheses caused a surge in attacks by climate science "skeptics," who argued that climate scientists were letting their personal values bias their research. Though an independent panel ultimately determined that the scientists had not manipulated data in ways that biased their research findings, the controversy damaged the credibility of the international climate science community (BBC News Online 2010).

The problem of dueling experts is exacerbated in the public health context by the cross-national and cross-sectoral nature of the issues. Each of the broad array of non-health-related stakeholders affected by a public health issue may seek out an expert to argue for the scientific evidence in favor of their position. The cross-sector linkages can also lead to clashes between rival epistemic communities— groups of experts with a deep knowledge of an issue area, such as infectious diseases, intellectual property, the pharmaceutical industry or primary health systems. These clashes are particularly hard to address because experts from different epistemic communities do not focus on the same set of factual questions. For example, one expert may focus on the best way to contain an infectious disease, while another concentrates on the likely impacts of travel restrictions on international trade. As a result, these experts may "talk past" each other while claiming to have the most relevant expertise on the issue at hand.

As an example, in 2006 the Bush Administration announced a $15 billion emergency plan for combating HIV/AIDS in Africa that promoted abstinence until marriage as a primary approach to fighting the pandemic. The UN Special Envoy on AIDS, Stephen Lewis, criticized this program as actually undermining the efforts of African countries to fight the epidemic, claiming that abstinence programs had been shown not to work. In the ensuing public exchange of mutual criticisms, each side claimed a lack of evidence for the other's position while pointing to "evidence" of their own (BBC News Online 2006).

Fortunately, there is an alternative to the dueling experts scenario—*joint fact-finding*. Joint fact-finding is a process in which diverse stakeholders work together to define the technical and scientific questions to be asked, and then jointly identify and select qualified experts to assist the group as a whole in finding answers. The stakeholders, together with mutually agreed experts, proceed through several steps. They refine the factual questions; set the terms of reference for technical or scientific studies; monitor, and possibly participate in, the study process; and review and interpret the results. While most joint fact-finding is done during the pre-negotiation phase, the technique can be applied throughout a negotiation process whenever there is a need to establish a common set of facts.[1]

The diagram on the next page highlights the main steps in a joint fact-finding (JFF) process, with specific actions for each step. Each step has one or two main purposes:

Assess the need for JFF: The critical first step in any JFF process is to clarify the scientific and technical issues to be addressed, based on an understanding of stakeholders' main concerns and questions, and of their willingness to collaborate in exploring the issues. Normally a stakeholder, or an outside body with an interest in promoting collaboration, will play the role of convener—that is, a party who invites (and may seek to influence and persuade) other stakeholders to participate in a joint fact-finding effort. Often conveners will ask the help of a neutral party who has both process facilitation skills and technical understanding of the issues to conduct a stakeholder assessment. The assessor will seek to talk with all primary stakeholders to understand their interests and concerns overall, and to see whether and under what conditions they might be interested in participating in a joint fact-finding process. The assessor may provide a report back to the convener and all the stakeholders interviewed, with an assessment of the feasibility and desirability of proceeding with joint fact-finding, and with recommendations on how to structure the JFF process. It is of course ultimately up to the stakeholders themselves to decide whether and how to proceed (Fig. 3.1).

In conducting the assessment, it is very important to distinguish scientific and technical questions from conflicting interests and values. JFF is useful for helping stakeholders investigate empirical issues in a constructive way, but it is not a substitute for interest-based negotiation or for dialogue to address underlying value conflicts. For example, it may be possible for stakeholders to answer the question "how prevalent is HIV/AIDS in our region?" through JFF. On the other hand, JFF alone cannot answer the question, "who should have the lead responsibility for HIV/AIDS education?" when there are conflicting organizational interests that must be negotiated; nor can it answer the question: "what proportion of an international program budget should go to HIV/AIDS prevention and treatment, and how much to other infectious diseases?" Answering that question will require both interest-based negotiation among professionals and constituencies with

---

[1] For an overview of joint fact-finding, *see* Ehrmann and Stinson (1999).

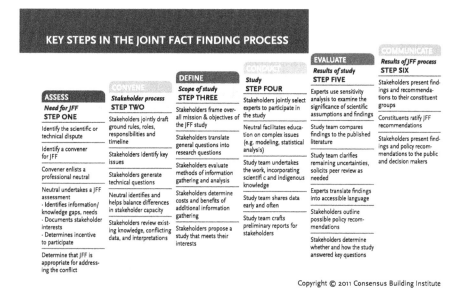

Fig. 3.1 Key steps in the joint fact finding process

different public health concerns, and dialogue about public values at stake in the allocation of scarce funding among different population groups.

It is equally important to assess the willingness of stakeholders to collaborate in exploring the issues. Stakeholders may be hesitant to collaborate on JFF for several reasons: because they believe that they already know the answers to the questions; because they believe that they can achieve their goals regardless of whether others agree with their view of the facts; because they do not believe others will participate in good faith, or because of a combination of these factors. The assessor needs to explore these concerns with hesitant stakeholders, in order to determine whether and how a joint fact-finding process could be designed to respond to stakeholder interests and concerns.

The assessor normally produces a report, in writing and/or as a face-to-face presentation to the convener and all stakeholders who have been interviewed. The assessment report should clarify the issues that could be addressed through joint fact-finding, the views of stakeholders on the potential for JFF to help them meet their goals, the options for proceeding with JFF and other means of addressing stakeholder concerns (e.g. direct negotiation). It should conclude with the assessor's own recommendations.

Convene the stakeholder process: If the stakeholders do agree to undertake joint fact-finding, the next step is to bring them together (convening them) to begin the JFF process. The key decisions to be made at this stage are to define the goals of the JFF process, the roles and responsibilities of participants, the core issues to be investigated, and the way that stakeholders will use information developed through JFF in their negotiation and decision-making. Stakeholders may also decide to use the services of a neutral facilitator to help them with the fact-finding process, and/or may ask the convener to provide ongoing technical and facilitation assistance.

In scoping the core issues, stakeholders should make use of existing studies and data and share their perspectives on the facts, with the understanding that existing information is not meant to resolve the major issues, but rather to clarify what information already exists, where there are gaps and disputes, and therefore where JFF is most needed to resolve outstanding questions. It is usually helpful to have a written statement of the goals and ground rules for a JFF process, and to establish a time frame for each JFF activity. A neutral facilitator can help the parties reach agreement on these critical "constitutional" issues as they begin the JFF process.

Define the scope of the study: This step takes the stakeholders further into the detail of determining the specific research/technical questions that need to be asked and answered, the methods to be used, and the experts/resource people who will do the work. For example, if stakeholders reached agreement that the core question for investigation was to determine "globally, how effective are abstinence and contraception methods for HIV/AIDS prevention for at-risk groups," they might then work together to define the methodology for reviewing the effectiveness of each prevention approach. For example, they might decide to rely only on a literature review of existing evidence, or to commission new clinical trials of each approach, or a hybrid with a literature review followed by clinical studies only to address questions not fully resolved by the literature review.

Conduct the study: After agreeing on questions and methods, the stakeholders need to select experts or resource people to help answer the questions. In some cases, stakeholders may be capable of conducting some of the investigation themselves, with agreement on ground rules to assure objectivity in their work. In other cases, stakeholders may need outside expertise and/or may lack trust in each other's ability to carry out study tasks objectively. Stakeholders then need to agree on outside experts to bring in. To avoid recreating the "dueling experts" problem in a joint fact finding exercise, it is useful for the stakeholders to agree on the criteria for expert selection first, then jointly evaluate experts and seek agreement on well-qualified individuals who are able to be impartial investigators of the questions.

During the study, stakeholders may ask for periodic reporting, or may simply wait for the results of the study to come back, and then discuss the results. The greater the complexity of the questions and the methods (for example, clinical trials are generally more complex and time consuming than a literature review), the more benefit there may be for stakeholders to discuss progress reports with the investigators. When stakeholders begin with highly uneven levels of technical understanding, periodic discussions with credible, impartial experts can help balance the level of technical sophistication among stakeholders. Conversely, through ongoing discussions with stakeholders, experts may learn more about the core questions and concerns that are driving the stakeholders, and may modify their approach to answering the questions in order to be as responsive as possible to stakeholder interests.

Evaluate: When the experts have completed their effort to answer the questions posed by the stakeholders, they need to make sure that their findings are credible both for the expert community and for the stakeholders and their constituencies. First, they need to clarify the explanatory power of their findings (whether statistically or through qualitative interpretation), and the sensitivity of findings to assumptions and conditions specific to the study design. Stakeholders and experts may also seek external peer review in some cases. Second, the experts need to communicate their findings in a way that is clear and accessible to the stakeholder representatives with whom they have been interacting directly, and to the constituencies that the stakeholders represent. For example, the findings of a study on the effectiveness of abstinence and contraception approaches to HIV/AIDS prevention may be of significant interest to a wide range of constituencies, to the media and the public at large. The experts will need to craft their report in language that makes the questions, methods, findings and interpretation as clear and accessible as possible to a broad audience.

However, expert interpretation is not the only or even the most important form of interpretation in a joint fact finding exercise. It is critically important that the experts have direct dialogue with the stakeholder representatives about their findings, and that the stakeholders themselves test and refine their own interpretations of the results jointly. Only through dialogue about the findings can the stakeholders gain the greatest benefit of a JFF process: a shared understanding of what is known and what is not on a complex factual issue. For example, the findings on abstinence and contraception approaches to HIV/AIDS prevention might show a significant variation in the effectiveness of each method, depending on the approach to information, education and communication with at-risk populations; the particular population in question; and the presence or absence of complementary public health interventions and services. To make the findings useful in resolving disagreement among the stakeholders, the experts and the stakeholders would need to have an extended discussion of the findings and their sensitivity to specific assumptions and conditions, and stakeholders would need to ask a number of "what if" and "what about" questions in order to fully probe the implications of the study.

Communicate: The final step in a JFF process is to communicate the results beyond the core group of stakeholder representatives, to their constituencies and the public. The initial expert presentation might be refined in response to stakeholder discussion, and stakeholders themselves may take direct responsibility for communicating the findings to their constituencies, either jointly with experts, or using written material produced by the experts. Likewise, stakeholders and experts might speak jointly to the media in order to present a shared understanding of the study results.

It is important to underscore that simply reaching agreement on the answers to a set of factual questions may not resolve all—or even the most important—of the issues that the stakeholders need to address. For example, knowing that promotion of contraceptive use is on average more effective than abstinence promotion for a particular age group may not resolve the question of whether it is morally acceptable to promote contraceptive use with that age group. The stakeholders may still face a difficult negotiation challenge in determining how best to use the results of joint fact-finding to resolve non-factual concerns.

Though it cannot resolve non-factual concerns, joint fact-finding by stakeholders with expert assistance does offer four major benefits. First, it enables parties in a negotiation to explore difficult topics together. Exploring the issues jointly allows stakeholders to develop a common knowledge base and an understanding of the "range of uncertainty"—the specific topics on which definitive factual answers do not exist. It also enables stakeholders to resolve disputes about technical and scientific methods, data, findings and interpretations before a negotiation begins. The amount of time and effort spent debating scientific issues during a negotiation can thus be dramatically reduced.

Box 3.1 Joint fact-finding efforts by trade and health officials can be particularly beneficial. Through this process, trade officials can gain a better understanding of the implications of strengthened patent protection, while health officials can become better equipped to discuss the economic costs and benefits for their countries of receiving better access to foreign markets. In contrast, where fact-finding is undertaken in a non-collaborative fashion, important perspectives may be missed. The latter occurred in a study by a trade ministry in Central America, which suggested that the short-term impact of increased patent protection would be limited, in particular, in relation to foreign market access benefits. Because health officials did not take part in the development of this report, it disregarded the long-term impact of increased patent protection on drug access and the policy options available to counteract the impact of the Central America Free Trade Agreement on drug prices (e.g., the parallel importation of drugs and the use of compulsory licenses) (Blouin 2007).

Second, joint fact-finding allows those stakeholders with less knowledge, education or expertise to learn more about the technical issues involved and the sort of data required at the international level. This enables negotiation on a more equal footing. For example, joint fact-finding on linkages between health and economic issues can be very helpful to health agencies in developing countries, generating good analyses that might not otherwise be available to them. In addition, representatives involved in international joint fact-finding may be better able to explain issues and policy options to leaders and key constituents in their home countries.

Experience has shown that taking the time before (or in) a negotiation process to develop a better technical understanding of the essential issues markedly improves the process and can in fact lead to better outcomes. The experiences of the Intergovernmental Working Group on Public Health, Innovation and Intellectual Property (IGWG) and the negotiations on virus sharing underscore the benefits of developing a joint understanding of technical issues, especially as diplomats without specific scientific knowledge often lead delegations. The IGWG was established in 2006 as an intergovernmental working group open to all Member States to draw up a "global strategy and plan of action" that would provide a "medium-term framework" and would, *inter alia*, aim to secure "an enhanced and sustainable basis for needs-driven, essential health research and development relevant to diseases that disproportionately affect developing countries" (World Health Assembly 2006). The initial negotiating session was relatively ineffective, as many of the delegations were confronted by complex issues not typically addressed by those in the public health realm. Many delegations came to the meeting without having done the necessary technical work and stakeholder consultations—the pre-negotiation preparation that would have facilitated a better outcome at this initial plenary session. A series of regional and inter-country meetings was subsequently organized to enable the national delegations to better understand the issues, dialogue with key stakeholders and develop negotiation options. When the delegations met a year later in plenary, the negotiating process was markedly improved.

Similarly, in the virus sharing negotiations in 2006 and 2007, the secretariat undertook to make detailed technical presentations on substantive issues concerning influenza and other health issues before and during each intergovernmental meeting. Diplomats highly appreciated these presentations on such technical issues as steps in the vaccine production, on which they were negotiating text—an event which often happens during a long and protracted negotiation process. The presentations facilitated a common understanding of the issues, and were particularly important for diplomats who were newcomers to the issues due to their recent change in postings and responsibilities.

The third benefit of joint fact-finding is that it facilitates greater creativity and better agreements. It enables parties to draw on each other's experience, knowledge and ideas, and can often result in innovative agreements that no single party could have generated alone. As there is no universally perfect method for collecting and analyzing evidence about health, and different circumstances call for different methods, the high level of collaboration enabled by joint fact-finding is vitally important. It results in a firmer technical and scientific foundation for later decisions or recommendations.

The fourth benefit is that joint fact-finding helps to improve relationships among parties with differing interests and perspectives. It enhances communication, fosters trust and helps build a deep understanding of others' interests, needs and values. Thus, it can bridge the gap between rival epistemic communities and between science and policy-making, thereby enabling more health-sensitive global policies. In addition, the shared investment of time, ideas and resources into jointly discovering good information increases the level of commitment among the parties to reaching a mutually agreeable outcome. The efficiency of a multi-stakeholder negotiation is further enhanced by minimizing the formation of adversarial coalitions supporting differing schools of thought.

In practice, joint fact-finding can take several forms:

• *Multilateral joint fact-finding institutions.* The Intergovernmental Panel on Climate Change (IPCC), the Millennium Assessment Process, and the Scientific Advisory Committee on Tobacco Product Regulation are recent examples of multilateral bodies created to conduct joint, integrated assessments that are technically credible and responsive to key policy questions.

Box 3.2 The IPCC was established in 1988 by the World Meteorological Association and the United Nations Environment Program and has recently completed its Fourth Assessment Report, *Climate Change* 2007. The IPCC offers many lessons, both for the increasingly important role health has played in each subsequent Assessment Report and the compositional changes of the assessment teams, which have over the years come to include more experts from developing countries and from a broader array of issue areas. The resulting ever-widening IPCC focus has slowly and hesitantly turned a mostly scientific, chemistry-focused enterprise into a more authentic, integrated assessment for sustainable development.

• *Non-institutional (more ad hoc) joint fact-finding processes convened by international organizations such as the WHO or by countries, to explore specific issues.* For example, throughout the negotiation of the FCTC, the WHO and a number of states convened technical conferences and consultations on topics ranging from "Potential Liability and Compensation Provision for the Framework Convention on Tobacco Control" (WHO 2001) to "Avoiding the Tobacco Epidemic in Women and Youth" (WHO 1999).
• *Joint fact-finding processes organized on a nongovernmental or quasi-governmental basis.* These processes—including entities such as health commissions— usually precede the initiation of official or institutional processes. One example is the Commission on Macroeconomics and Health, chaired by Dr. Jeffrey Sachs, which demonstrated and quantified how health contributes to economic growth; it found that "health status seems to explain an important part of the difference in economic growth rates [among countries], even after controlling for standard macroeconomic variables," and thus characterized health as a good investment (Sachs 2001, p. 24). Another is the Commission on Intellectual Property Rights, Innovation and Public Health, which, looked at how to move research and development funds into diseases that affect the poor. It found that where the market does not work, mechanisms other than intellectual property rights are needed in order to provide the incentives for research and development. Ultimately, the commission process resulted in a global strategy for public health innovation and intellectual property. Another commission—the Commission on Social Determinants of Health, which had extensive knowledge networks of experts, showed that where one lives, plays and works are important determinants of health outcomes and need to be integrated into the development

of strategies for better health. This commission brought equity to the forefront; it showed that where one lives and plays is as important as the health care system in the community in which one lives.

3.2 Developing Options and Packages: The Mutual Gains Approach

Once the issues have been framed and some shared understanding of the facts has been established, the actual negotiation stage begins. The conventional view of complex international negotiations is that they are necessarily conflictual, with stakeholders battling to achieve incompatible goals. As an example, developing countries may seek to gain low-cost access to medicines produced in developed countries, challenging intellectual property rights that allow title-holders to charge prices above marginal costs. Developed countries, on the other hand, may seek to protect their domestic drug manufacturers and insist on the implementation of property rights legislation as mandated by TRIPS (Correa 2006). The magnitude of global public health challenges such as these demands the development and use of truly effective negotiation strategies.

The conventional strategy in these kinds of cases, unfortunately, is hard bargaining. In hard bargaining, parties set out extreme positions, withhold information and make concessions grudgingly. Interpersonal interactions may be difficult, especially when the representatives or their organizations have a history of conflict, or when they are skeptical of each other's commitment to a good-faith negotiation process.

The problem with a hard-bargaining approach is that it assumes that interests are incompatible and mutually exclusive. In reality, most multi-party, multi-issue international negotiations have at least some potential for all stakeholders to make gains relative to the *status quo*. In the public health context, the presence of many stakeholders and many issues generally allows for substantial joint gains among most stakeholders on some, if not all, issues. Hard bargaining fails to realize those gains, however, and settlements are less likely to be economically, environmentally or socially sustainable.

Potential joint gains may likewise be left unachieved where a "soft bargaining" strategy is adopted. In this approach, negotiators avoid contentiousness at all costs and sacrifice their own interests in order to reach agreement and maintain good interpersonal, organizational or inter-state relations. Or they may give into a more powerful party in the hopes of gaining something (or avoiding negative action) in another domain such as trade or development assistance.

In most cases, the most efficient and sustainable negotiation outcomes can be achieved by seeking to meet one's own interests *and* those of one's counterparts, thus preserving and improving ongoing relationships with other negotiators and the organizations they represent. The *mutual gains approach* to negotiation, developed at the Program on Negotiation at Harvard Law School, is a strategy for achieving efficient and sustainable negotiation outcomes in this manner.

Applying the mutual gains approach can greatly improve one's capacity to meet public health goals. It offers strategies for each of the four stages of the negotiation process:

1. Preparation (before the negotiation)
2. Value creation (initial stages of developing options that are advantageous for all sides)
3. Value distribution (reaching agreement)
4. Follow-through (implementation)

Box 3.3 The key principles of the mutual gains approach are:

- Prepare effectively by focusing on stakeholders' interests and best alternatives to a negotiated agreement (BATNAs)[2] and by generating initial proposals for mutual gains.
- In value creation, begin by exploring needs and interests, not by stating positions.
- To find potential mutual gains, use no-commitment brainstorming to develop options and proposals that might meet both one's own needs and interests *and* those of other stakeholders.
- Seek maximum joint gains before moving to value distribution (i.e., making commitments and deciding "who gets what").
- When distributing value, find mutually acceptable criteria for dividing joint gains.
- In follow-through, ensure that agreements will be sustainable by committing to continuing communication, joint monitoring, contingency planning and dispute resolution mechanisms.

The following sections describe and offer advice regarding the first three steps of the mutual gains process, in the global public health context; step four, follow-through, will be discussed in Chap. 6, Meeting Implementation Challenges.

Figure 3.2 provides an overview of the mutual gains approach.

---

[2] Best Alternative to Negotiated Agreement (BATNA) is a term of art popularized in Fisher, Ury and Patton (1991). It refers to what a negotiating party will do or can get *away* from the negotiating table, without the agreement of the other side. It is his/her *alternative* to agreement with the other side.

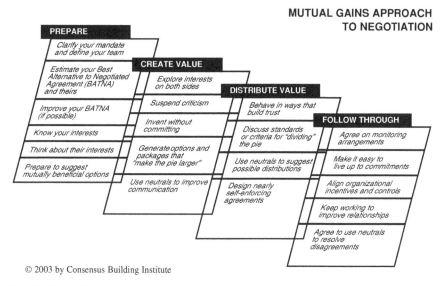

Fig. 3.2 Mutual gains approach to negotiation

3.2.1 Getting to the Table: Preparation

The first step in the mutual gains approach is preparation for the negotiation. At the core of preparation is a careful, dispassionate analysis of the relevant parties, their goals and interests and their alternatives to a negotiated agreement. It is very important to know one's own interests, one's alternatives if the negotiation fails to produce agreement (i.e. one's "no negotiation" alternatives, or BATNA) and one's minimum acceptable conditions for the agreement ("bottom line") based on an assessment of those alternatives. When representing others, preparation together with constituents, colleagues and/or leaders is critical. Finally, in order to be able to develop options for mutual gain, it is *equally essential* to understand the interests, alternatives and bottom lines of one's negotiation partners.

Good preparation has both substantive and psychological benefits. Substantively, well-prepared negotiators maximize their ability to get what they want. Psychologically, they can keep cool during the negotiation process, be creative and be helpful to their negotiating partners—without going to extremes of "giving in" or "playing hardball."

The Alliance of Small Island States' (AOSIS) participation in multilateral climate change negotiations illustrates the benefits of good preparation. AOSIS negotiators were highly effective in making their countries' concerns heard and having their national interests met, because they rigorously prepared for each negotiation session and developed briefing books for AOSIS members. They focused on helping each other identify their strengths and capitalize on their resources, so that they would be in at least as good a position as any other developing country delegation. For this purpose, they frequently brought in experts on specific topics to brief them in detail. Moreover, they analyzed their negotiating partners' interests and constraints, especially developed country parties, and developed options and arguments that responded to them. Thus fully prepared to engage in substantive discussions, the AOSIS representatives gained political influence and were not ignored. Instead, thorough preparation enhanced their credibility in the negotiation process (see Chap. 9).

Developing countries were similarly prepared and played an equally influential role in negotiating the Doha Declaration on TRIPS and Public Health, which was approved by trade ministers at the Fourth WTO Ministerial Conference in Doha, Qatar, on November 14, 2001. The Declaration was the product of months of negotiations that examined TRIPS and its impact on the public health sector. Developing countries largely achieved their objectives in these negotiations because of their level of preparation. They submitted several official written proposals and were clearly abreast of important concepts behind the major issues to be resolved. As early as the TRIPS Council's first special session on access to medicines, the Africa Group, side-by-side with many other developing countries, issued a paper on its view of the relationship between TRIPS and access to medicines. It also introduced a set of limiting principles on the procedural aspects of the negotiation to follow. The paper consistently cited the applicable portions of TRIPS and offered effective interpretations of TRIPS. Through their strong understanding of the legal foundations of the issues in the negotiation, the developing countries were able to make effective arguments in support of their interpretation of TRIPS.[3]

3.2.1.1 Distinguishing Interests From Positions

Central to effective preparation is the analysis of negotiating parties' interests. The mutual gains approach focuses on interests rather than positions. A *position* is the stance a party takes on an issue (e.g., "we are going to allow domestic drug producers to manufacture generic drugs without first obtaining licenses" or "licenses must be given by patent holders for production of generic drugs"). It is what a party is demanding, its advocated solution. In contrast, an *interest* is the core need, want, fear or concern that underlies a position and forms the reason(s) and goal(s) behind the position—*why* the party wants its position (e.g., "we need access to affordable, life-saving drugs for our large and impoverished population," or "we fear that companies will not invest in research and development of new and important medicines").

---

[3] *See* "Analyzing a Complex Multilateral Negotiation: The TRIPS Public Health Negotiation," Chap. 7 in this volume.

Box 3.4
Distinguishing Interests From Positions
Position: What you want
Interest: Why you want it
Focusing on Interests in Negotiations

- In preparation, analyze both your interests and their interests
- At the table, explain your interests
- Ask questions and listen to discover their interests

Negotiators' interests may include, for example, protecting public health, promoting development, making a profit, satisfying shareholders, enhancing organizational reputation and image, generating resources to pursue their missions, improving relationships with key counterparts, establishing precedents for future negotiations or gaining fair treatment on an issue, among many others.

By focusing on interests rather than positions, negotiators can open up new possibilities for mutual gains or a way out of a deadlock. A position is one way to meet an underlying interest, and is often presented as a "take it or leave it" choice. In contrast, an interest may be met in any number of ways, and it does not have to be presented as a demand or ultimatum. Often, the discussion of interests can open up space for brainstorming options—i.e., ways to meet the interests of the participating stakeholders—while the presentation of positions can leave negotiators feeling as if they have little to discuss.

To prepare effectively for negotiation, it is essential to clarify one's own interests. One useful way to distinguish interests from positions is to state something that might be an interest, and then ask oneself: "Why do we want that?" "What do we want that for?" "Why is that important to us?" If the answer is a more general way to achieve the same goal, yet something that might still be negotiable, the negotiator has made progress toward more clearly defining his or her interests. For example, a negotiator might first state his or her interest as "increasing staffing in our primary health clinics." After asking "why?" the negotiator might answer, "to improve primary health care service delivery." The second statement may be a more useful framing of the interest, because there may be alternatives to increased staffing in the primary health clinics that could be equally or more effective in improving primary health service delivery. A good test of whether one is really getting to interests is whether there is more than one solution to meet the interest; if the stated "interest" leads to only one solution, then a negotiator should continue to ask, "why?"

In an organizational context, negotiators should define their interests through dialogue with those whom they will be representing, be they senior managers, colleagues and/or constituents. Jointly answering the "why" questions should help clarify organizational goals, and may also be a good way to identify trade-offs or competing interests within the organization. However, the process of defining organizational interests in a negotiation does not end when the representative goes to the negotiation table. On the contrary, effective negotiators maintain regular communication with those they represent during the negotiation process, to summarize the status of negotiations, test possible options and trade-offs and reassess interests in light of new information and ideas.

As negotiators define their own interests, it is equally important that they assess the likely interests of their negotiation partners. This is often easier when a negotiation takes place between organizations and individuals who have worked together before, understand each other well and understand the issues well. It may be much more difficult when the negotiators and their organizations do not know or understand each other, where the issues and options are not entirely clear, or where there are clearly different or opposing positions. The negotiation of health issues involving the WHO secretariat in Geneva, Switzerland, for example, often involves health attachés from country missions who are posted in Geneva. The attachés' frequent formal and informal contact among themselves and with the WHO secretariat responsible for convening and servicing the negotiations builds a certain level of trust and understanding that can greatly facilitate identification of underlying interests.

In the Intergovernmental Working Group on Public Health, Innovation and Intellectual Property (IGWG), for example, the initial positions of countries presented during the first plenary session reflected differing views on intellectual property and its effects on innovation and access. Developing countries rejected intellectual property rules, while the pharmaceutical industry insisted on preservation of the existing patent system in relation to development of drugs, including for "diseases of the poor." Little progress was made until regional and country consultations were held, and countries began to understand the issues, their own interests and other countries' interests. While positions were rigid, interests were less opposed; developing countries were concerned about the development of drugs for diseases of the poor, while industry did not want intellectual property to be undermined in areas where the market worked. This understanding of the interests of the parties revealed possibilities for options regarding pharmaceutical development for "diseases of the poor" that could satisfy both sides. Part of the agreed strategy and action plan was to come with innovative financing mechanisms for diseases of the poor.[4]

---

[4] *See* The Global Strategy and Plan of Action on Public Health, Innovation and Intellectual Property (GSPOA), WHA 61.21 (61st World Health Assembly, 24 May 2008) and WHA 62.16 (62nd World Health Assembly, 22 May 2009), http://www.who.int/phi/implementation/phi_globstat_action/en/index.html. *See also* Expert Working Group on R&D financing and coordination. The Expert Working Group was formed as a "results-oriented and time-limited Expert Working Group" in response to the World Health Assembly Resolution on the Global strategy and plan of action on public health, innovation and intellectual property to "examine current financing and coordination of research and development, as well as proposals for new and innovative sources of funding to stimulate research and development related to Type II and Type III diseases and the specific R&D needs of developing countries in relation to Type I diseases." (Expert Working Group on R&D financing 2010, http://www.who.int/phi/ewg/en/index.html).

Similarly, in negotiations over virus sharing that resulted in the adoption of WHA60.28 (2007), Indonesia initially refused to share samples of the H5N1 (avian flu) virus, claiming that under the Convention on Biological Diversity, access to and use of virus samples obtained in Indonesia could occur only with Indonesia's consent. Led by the United States, other countries insisted that Indonesia was obliged to share data and virus samples without preconditions, and that Indonesia was in violation of the International Health Regulations 2005. A focus on the underlying interests permitted a resolution under which Indonesia resumed sharing of the H5N1 virus. Indonesia's interests were in gaining benefits from the knowledge and technologies derived from use of the samples, while the interest of the US, WHO and others was in containing the spread of the virus, and, more generally, in preserving and strengthening the global surveillance system and the development of intervention strategies to deal with such epidemics. The WHA resolution did not resolve the legal question of whether unconditional virus sharing was required by the IHR 2005; instead, it outlined processes for agreeing to terms and conditions for sharing of viruses between the originating countries, WHO Collaborating Centres and third parties, as well as for ensuring resulting fair and equitable sharing of benefits, while directing the WHO Director General to establish an international stockpile of vaccines "for use in countries in need in a timely manner and according to sound public-health principles, with transparent rules and procedures".[5]

The greater the uncertainty or difference of views regarding the negotiating topic, the greater the potential benefit of "doing one's homework" on the interests of counterpart organizations and their representatives. It is important to consider: What is their overall set of goals? What involvement have they had in this issue to date? What publicly available statements, papers, news reports or other documentation outline their views on the issue? Does anyone in one's own organization have a personal connection to the counterpart organizations, from whom the negotiator could hear some of the counterpart's perspectives on the issues? Doing some work to answer these questions will increase the potential for joint gains in the negotiation process, and will help to identify areas of potential conflict. Negotiators can use a relatively simple worksheet like the one on the next page to organize their thinking in preparation.[6]

With a clear understanding of one's own interests, and good information or well-educated guesses about the interests of others, the negotiator has taken the first solid step on the road to effective preparation.

---

[5] World Health Assembly Resolution 60.28, para 2.2 (23 May 2007).

[6] *See also* the Negotiation Preparation Worksheet in Appendix 2.

Negotiation Preparation Worksheet

	Ourselves	Stakeholder A	Stakeholder B	Stakeholder C
Interests?				
BATNAs?				
Questions to ask others?				
Options that are good for us and acceptable to others?	Issue 1 Issue 2 Issue 3			
What makes our preferred options fair?				
Implementation challenges and ways to address?				
Other issues/ strategies				

3.2.1.2 Assessing and Defining Aspirations, BATNAs and Minimum Requirements

Knowing interests is necessary but not sufficient for effective preparation. It is also important to define the spectrum of acceptable outcomes for oneself and consider what that spectrum might look like for the other negotiators. Acceptable outcomes may range from minimum requirements (the least that a negotiator can accept) to aspirations (the best possible outcome to satisfy a negotiator's interests).

To define aspirations, one must have a clear conception of one's own preferences and a clear enough sense of the interests of others to ensure that one's own aspirations will not be perceived as "nonstarters." For example, if one's interest is to develop an effective regional infectious disease surveillance program, one's aspiration might be to establish a new standing body to carry out the surveillance program, with full participation by all of the health ministries, funded by regional and global health donors. That aspiration might be very appropriate if all the health ministries in the region are willing in principle to participate, and the potential donors have signaled interest in funding such a body. If it is clear that some governments are not keen to participate, however, or that there is little donor interest, the negotiator should ask whether the aspiration is overly ambitious, and if so, reformulate it. A redefined aspiration might be to gain agreement among a core group of health ministries to pilot a self-funded surveillance system, and to seek full regional coverage and external funding after the pilot phase.

Having defined his or her interests and set an aspiration, the negotiator should decide the minimum outcome (i.e., the bottom line) that he or she would be willing to accept in an agreement. To determine one's bottom line, it is essential to consider what alternatives are realistically available to advance one's interests if an agreement cannot be reached. One's "best alternative to a negotiated agreement," or BATNA, should be the basis for calculating the bottom line.

A BATNA is *not* the same as a bottom line. One's BATNA is the best alternative for meeting one's interests away from the negotiating table, if it is not possible to reach agreement with one's negotiating counterpart(s)—a "plan B" in case negotiations do not produce anything that meets one's most important interests. In the surveillance example, the negotiator's BATNA might be to strengthen his or her own country's partnership with a regional health organization to provide the best available data on regional infectious disease incidence to his or her Ministry of Health. This alternative would not require the agreement of the other countries at the negotiating table nor additional donor funding.

Once the negotiator is clear on his or her BATNA, the next step is to define the minimum acceptable outcome of the negotiation—the bottom-line outcome that is just slightly superior to one's own BATNA. If the best possible proposed agreement (resulting from a good faith effort to create joint gains) is not better than the BATNA of the negotiator, then he or she should say "no thanks" to the proposed agreement and go with the BATNA. Again using the surveillance example, if only one other country in the region is interested in a joint approach to surveillance, and harmonizing procedures and communications with that country would be more costly than making more use of the data already collected by a regional health organization, then the negotiator should say "no thanks" to the bilateral surveillance agreement and proceed to strengthen communications with the regional organization. On the other hand, a deal with two out of three neighboring countries, including the country that poses the highest cross-border infectious disease risks to one's own country, might be good enough to beat the negotiator's BATNA.

From this presentation of BATNA analysis, it should be clear that a negotiator who prepares well never accepts a deal that is not better than his or her BATNA. A good BATNA is a significant source of bargaining power. For example, if one is negotiating a pharmaceutical licensing agreement with a company, and another pharmaceutical company with a comparable product is also very interested in reaching a licensing agreement, one has a good BATNA in the negotiations with the first company. On the other hand, if the drug in question is critical to public health and only one company is offering to license it, then one's BATNA may not be very good, and one may need to work especially hard in the negotiation process with that company to generate a mutually acceptable outcome.

Of course, assessing the BATNAs of one's negotiation partners is also absolutely critical to effective preparation. Having a good sense of the BATNAs of others should give the negotiator a clear idea of how far he or she needs to go to meet their interests in the negotiation. If negotiation partners have very good BATNAs, then one's own proposals will have to meet their interests very well. If one's negotiation partners do not have good BATNAs, then one's own proposals can offer less to them and claim more value for oneself. However, the effective mutual gains negotiator does not focus primarily on getting as much as possible for oneself at the expense of others, even if others have weak BATNAs. On the contrary, the effective mutual gains negotiator seeks agreements that meet the interests of all parties as well as possible, in order to create strong commitment to implementing the agreement and realizing joint gains.

The negotiator may have to do quite a bit of detective work to understand the BATNAs of negotiating partners. It even may require educated guesswork. Generally it is not in the interests of negotiators to reveal their BATNAs, and asking too directly about the BATNAs of others may raise questions about one's commitment to engage in good-faith negotiations. It may be possible to shed light on others' BATNAs by doing research that does not require direct contact with the potential negotiating partners. For example, researching a pharmaceutical company's current licensing arrangements and opportunities—based on data available on the Web, from business analysts and from countries that have licensing agreements with the company—may help the negotiator form a clearer sense of the company's likely BATNA (in this case, perhaps potential licensing agreements with other countries). Even if it is not possible to generate a very clear picture of negotiating partners' BATNAs, it is important to try, in order to put one's own negotiation strategy on as firm a foundation as possible, and also to recognize areas of uncertainty and identify questions to pursue during the negotiation process.

Whenever possible, a negotiator preparing to come to the table should not only assess and understand his or her current BATNA, and that of the other party(ies), but should also seek to *improve* his or her own, and *worsen* the other's. For example, if one were preparing for a licensing negotiation with a pharmaceutical company, it would be highly advantageous to see whether any other companies have a comparable product, and if so whether they are interested in licensing the product, before sitting down with the first company.

The benefit of investing time and effort in improving one's BATNA is illustrated by the case of Brazil's negotiation for access to HIV/AIDS medicines. The bilateral dispute between Brazil and the United States over Brazil's protection of intellectual property gained momentum when Brazil introduced a program of fighting AIDS and changed domestic legislation to facilitate its implementation, including permitting local manufacture of HIV/AIDS drugs. The U.S. believed the new program directly violated Brazil's obligation to protect intellectual property rights under the TRIPS agreement. Brazil, on the other hand, maintained that it had the right to use all necessary means to save its population from the AIDS pandemic. Among the most effective tools used by Brazil in these negotiations was the development of a very good BATNA—providing a framework for local production of drugs, as well as supporting local manufacturers and building coalitions with other developing countries with a strong pharmaceutical sector. Brazil also worked effectively on strategies to weaken the United States' alternatives to a negotiated agreement, enhancing its bargaining power significantly. Brazil took advantage of available assets, such as international law, domestic intellectual capital and the fallback option of legally issuing compulsory licenses for AIDS drugs and having the capacity to proceed with local manufacturing in case negotiations failed (see Chap. 8).

3.2.1.3 Preparing Options and Proposals for Joint Gain

Once a negotiator clarifies his or her own interests, BATNAs and minimum requirements, and forms well-educated guesses on those of his or her negotiating partners, it is time to formulate some options and proposals to bring to the negotiating table. A well-prepared negotiator comes to the table with one or more options and proposals that would meet one's own interests very well and are likely to meet the interests of other negotiators well enough to become the basis for further discussion. Each option should demonstrate a solid grasp of the issues and their technical, financial and institutional context; an understanding of the interests of negotiating partners; and one's own commitment to find agreements that are good for all or nearly all of the negotiators, not only for oneself.

In preparation, the basic question the negotiator needs to answer is, "Given what I know of my own interests, the interests of other negotiators, our BATNAs, the set of technically feasible options for meeting our interests and the resources that appear to be available to us, what could I propose that would meet my own interests well and would also be attractive to other negotiators?" Answering this question may be simple. In the case of a pricing negotiation for a drug with a well-defined market, there may be a well-established price, and the negotiation may turn on whether and how much discounting is feasible or how one might avoid setting a precedent for negotiations with other countries. Alternatively, answering this question may be extremely complicated. In negotiating a global strategy on intellectual property issues or a treaty on tobacco control, hundreds of actors and dozens of issues may be involved, and creating a proposal that can meet the interests of each actor on every issue may not be the most efficient way to prepare. Instead, the well-prepared negotiator:

- Considers the core issues that would have to be addressed in an agreement;
- Identifies the key parties whose agreement on the core issues could catalyze broader agreements with other actors on other issues;
- Develops one or two options on each of the core issues; and
- Considers what might need to go into a package agreement across all the issues in order to gain the support of the key parties.

Whether simple or difficult, preparing options to bring to the table is an extremely important part of any negotiator's preparation. In a sense, a negotiator who prepares options carefully anticipates the whole negotiation. He or she can gain a great deal of leverage in the negotiation process by identifying potential areas of agreement, differences that will need to be bridged to "create value" once the negotiators are at the table and options that might serve as bridges to agreement.

The negotiator should keep in mind one caveat as he or she moves from the preparation phase to the negotiating table: the options generated during the preparation phase should not become positions. They are ideas meant to jump start the value creation process. The negotiator should retain an open mind with regard to the efficacy of these ideas, as they will be more fully informed once the negotiator reaches the negotiating table and encounters his/her counterpart(s).

3.2.2 "Enlarging the Pie:" Value Creation at the Negotiating Table

Positional bargaining often results in "lowest common denominator" agreements or agreements about which all the parties are equally unhappy—if any agreement is reached at all. By contrast, the mutual gains approach challenges parties to "enlarge the pie"—to create as much value as possible for all stakeholders in the initial stages of a negotiation, before deciding "who gets what" in a final agreement. *Value creation* means inventing options that meet parties' interests well—meaning that the options are significantly better for all negotiators than their BATNAs and ideally are closer to their aspirations than to their bottom lines.

The pre-negotiation preparation steps just discussed are critical in enabling effective value creation. But they are not enough. The dynamic interaction of negotiators in a face-to-face setting has a profound effect on whether mutual gains can be realized at the table. It may generate new information, ideas, obstacles and options that no negotiator could fully anticipate during the preparation process. And if the interaction involves exchanges of positions, proposals and counter-proposals, or is adversarial, value creation may be undermined, and even options that are advantageous for all parties may be rejected. It is therefore important for the negotiator to have a strategy and tools for making face-to-face negotiations as constructive as possible. The mutual gains negotiator has two particularly powerful strategies, each with a simple tool. First, *clarify interests* by asking and answering "why?" questions. Second, *create and refine options for joint gain* by asking "What if?" questions; these allow negotiators to brainstorm ideas without saying that they will necessarily agree to any of the options under discussion. We will discuss each of these in turn.

The well-prepared negotiator comes to the table with a clear sense of his or her own interests, a good guess as to the interests of other parties. A good negotiator must also explore interests with the other side directly to make sure he or she understands them well before moving on to propose options. Asking questions of other negotiators is the best way to jointly explore and clarify others' interests.

Questions should be asked sincerely, not as a debating tactic and not as a way to undermine other negotiators. For example, a mutual gains negotiator will not ask, "Why are you being so vague?" or "Why can't you simply agree to what seems like an obvious point?" Instead, a mutual gains negotiator will ask, "Can you clarify that for me? I'm not sure what you meant..." or "I'm having a hard time understanding your concern on that point. Do you have a different understanding of the facts, or do you share my understanding of the facts but disagree about what I've proposed we do in response?"

Box 3.5 What it sounds like to explore interests:
- "What are the key things you need from an agreement?"
- "Why is that important to you?"
- "What else is important to you?"
- "Would you prefer [X] or [Y]?"
- "Could you live equally with [option X] and [option Y]? What do you like about the options?"
- "You've mentioned [X] and [Y] and [Z] as things that matter to you. Among these, which is most important?"
- "What concerns you about this proposal?"

With a solid understanding of interests, negotiators can then explore multiple options for resolution or collaboration using "what if" questions: "What if we tried a different option that could work for me, and if I understand your interests correctly, could work for you too?" or "What if we tried an option along these lines—would this be moving in the right direction?"

In order to find options that potentially meet all parties' needs, rather than make offers, it is useful first to suspend judgment about ideas that are raised and to invent options without making substantive commitments or even attributing ideas. The more options parties can come up with, the more likely they are to find something that will work for themselves and others. The "what if" technique avoids locking parties into their preconceived positions and ideas before all potential options have been explored.

Brainstorming without committing—in other words, without accepting or rejecting options—is difficult. The creative brainstorming process may be facilitated by involving representatives—such as mission staff in Geneva—who are explicitly *not* authorized to make decisions, at least at that point in the negotiation. Such informal processes can be part of the negotiation itself. In the FCTC process, for example, six important and difficult issues were discussed in informal meetings during the fifth session of the Intergovernmental Negotiating Body. At the sixth and final session, two informal groups were created to discuss the topics that were still causing hesitation among representatives: financial resources (which some developing countries said they required in order to comply with the convention) and advertising/promotion (which some developed countries were hesitant to restrict). These sessions led to the development of options that facilitated conclusion of an agreement on the FCTC text.

Box 3.6 Rules for Brainstorming
1. **Invent without committing:** brainstorming proposals are not formal offers. They can be discussed, but cannot be accepted or rejected
2. **The more options, the better**: be creative and come up with many ideas, even if some of what you invent is not workable
3. **The test of options is how well they might meet interests**: try to invent options that would be good for all parties; and focus discussion on how an option might be improved to meet more interests or meet interests better
4. **Work on the problem together**: in a brainstorming session, encourage negotiators to express their interests and concerns clearly, but all nego- tiators should be working together to find new ideas and options, not on defending their positions or their preferred options

---

The brainstorming process may, alternatively, occur outside the negotiating sessions themselves, or between sessions, amongst representatives in Geneva, for example, or in consultations with the chair or secretariat, as occurred in the FCTC process as well. This underscores the importance of a mission's involvement in the pre-negotiation and negotiation process. Countries who do not have missions in Geneva, as well as small missions that cannot devote time to all issues, can be at a disadvantage in this process. However, they can overcome this disadvantage in part by participating in regional groupings and strategic alliances, as well as pushing for and participating in regional consultations organized by WHO or other international organizations.

Whatever approach is used, the basic point is to avoid locking into positions by creating an atmosphere of joint problem-solving rather than hard bargaining negotiation. Nevertheless, there are situations in which a bad history and/or negotiating strategy make it hard for the parties to talk to each other openly. In these cases, an impartial third party can help the parties to communicate and develop options. During the Pandemic Influenza Preparedness negotiations, for example, seven countries involved in the Oslo Foreign Policy and Global Health Initiative convened pre-negotiation sessions aimed to let key delegations better understand each others' positions on elements of the negotiation text. This facil- itated agreement on certain parts of this text in subsequent formal negotiation sessions.[7] Similarly, during the end game of the IGWG in Public Health, Inno- vation and Intellectual Property negotiations, through a 'friends of the chair' group the chair facilitated the final tradeoffs that were needed to successfully conclude the negotiations on the global strategy. Here, as in many multilateral negotiations, the "third party" consisted of parties to the negotiation who had a strong interest in

---

[7] For more information on the Intergovernmental Meeting on Pandemic Influenza Preparedness, *see* http://apps.who.int/gb/pip/e/E_Pip_oewg.html. The Intergovernmental Meeting requested the Director General of WHO to convene an open-ended working group (OEWG) to continue working on this issue.

achieving an agreement and whose substantive interests were sufficiently represented by other delegations for them not to have to advocate for them.

In the climate change negotiations on emissions reduction targets for industrialized countries (leading ultimately to the creation of the Kyoto Protocol to the
Framework Convention on Climate Change), a number of think tanks and conflict
resolution organizations organized informal brainstorming sessions on difficult
issues. Most of these sessions involved a mix of government negotiators, experts
and advocates from environmental groups and industries. Some were facilitated by
professional facilitators, others by experts with reputations for impartiality. Sessions ranged from a few hours to a few days in length. Several of these sessions
generated useful ideas that participants fed back into the formal negotiation process, including the idea that eventually led to the creation of the Clean Development Mechanism (Martinez and Susskind 2000).

3.2.3 Reaching Agreement: Value Distribution

Once negotiators have generated good options that all or nearly all parties see as
potentially better than their BATNAs, they may still face hard choices as they seek
to finalize an agreement by choosing among the options identified. The third step
of the mutual gains approach, *value distribution*, offers strategies for achieving
joint gains while dealing with the reality that not all parties will be equally satisfied
with any proposed agreement. The danger for negotiators is that the struggle to
"get the biggest piece of the pie" they have created will undermine the potential
for reaching a mutually beneficial agreement.[8]

To reduce the risk of deadlock, the mutual gains negotiator first seeks an
agreement on "objective" principles, standards or criteria for choosing among
options, instead of resorting to hard bargaining. "Objective" does not mean
"right," but rather, acceptable to all parties as a reasonable and fair way to make a
decision, and not just as a cover to justify individual preferences. Such criteria in
the public health context might include the probability of reducing infection rates,

---

[8] For example, a potential home buyer may be willing to pay more than the minimum amount a
seller would be willing to accept. If both sides use hard-bargaining ploys, however—with the
seller claiming he will not accept less than the initial asking price, and the buyer claiming she will
not pay more than her initial offer—they may not reach a deal, even though a mutually beneficial
price does exist that is significantly better than their bottom lines. The mutual gains prescription
here would be for each to assess their BATNA to see whether further negotiation was in their
interest. Assuming it is (i.e., there is not a more attractive house available to the buyer, and the
seller does not have or expect a better offer), the two negotiators should look for criteria, such as
the prices of comparable homes in the same market, that both would agree are a fair basis on
which to set the price. It might also be possible for the buyer and seller to introduce new options
on issues other than price, such as the time needed to complete the sale, or the completion of
repair work, that could make it easier to reach a "package agreement" that buyer and seller would
both consider fair.

morbidity and/or mortality; program cost-effectiveness; impact on incentives for research and development; equity in cost sharing; and administrative feasibility, among many others.

It is often useful to develop agreement on guiding principles, standards or criteria at the beginning of a negotiation process. Doing so can create a sense of shared purpose and mutual understanding among representatives in a negotiation, prior to a detailed exploration of the facts and the development of options. And there is an additional benefit: At the outset, negotiators do not yet know for certain what options will be developed during the negotiation, and so they are less likely to engage in hard bargaining for criteria that are very narrowly targeted to their preferred options. For example, it may be easier to reach agreement on the criterion "highest probability of reducing infection rates" at the outset of a negotiation, when there is still substantial uncertainty about which approach might best reduce infection rates. If the discussion of criteria happens after joint fact-finding and option development have produced two options for a health intervention, each strongly supported by a different set of parties, and each with a different probability of reducing infection rates, then discussion of the criterion "higher probability of reducing infection rates" is likely to be colored by the conflicting interests of the two sets of parties.

In the negotiations to create the Framework Convention on Climate Change, one of the most challenging questions was how to create a fair balance of responsibility between developed and developing countries for responding to climate change. This issue almost deadlocked the negotiations several times. Developed countries recognized their historic responsibility for fossil fuel burning, but were concerned that fast-industrializing developing countries' emissions were now growing much faster than their own. Developing countries took the view that developed countries had created the problem and should take responsibility for correcting it, without constraining developing countries' industrialization paths. Skillful negotiators from developed and developing countries ultimately worked out a principle of "common but differentiated responsibility" for reducing the risk of climate change:

> The Parties should protect the climate system for the benefit of present and future generations of humankind, on the basis of equity and in accordance with their common but differentiated responsibilities and respective capabilities. Accordingly, the developed country Parties should take the lead in combating climate change and the adverse effects thereof (Framework Convention on Climate Change 1992, Article 3, paragraph 1).

The Convention gave shape to this principle by committing developed countries to begin reducing their emissions first, and to provide financing and technology to developing countries to enable them to reduce the rate of growth of their emissions without compromising their economic development prospects.[9] However, the principle and the general commitments laid out in the Convention did not tightly constrain the actions to be taken by developed and developing countries, allowing them latitude to work out details through further negotiation.

---

[9] For a multi-participant negotiation history, see Minzter and Leonard (1994).

Often it is not possible to discuss and agree on criteria at the beginning of a negotiation. Negotiators can still avoid hard bargaining among competing proposals during the *value distribution* phase of the negotiation by asking questions about the criteria on which the proposals are based. Particularly for negotiators from developing countries, insisting on and discussing appropriate objective criteria for choosing among options can be a powerful tool to resist coercion or pressure from more powerful negotiating counterparts.

Box 3.7 What it sounds like to explore criteria:
- "How did you arrive at that?"
- "What makes that fair?"
- "How can I justify this to my people?"
- "What kind of argument would your people need to hear to support this?"
- "How are others (people, organizations) handling this problem?"

If the parties themselves cannot come up with mutually acceptable criteria, they might present their preferred outcomes to a neutral third party they trust for input and/or a decision—for example, someone from the private sector or civil society sector with experience with the issue at hand.

3.2.3.1 Resolving Disagreement through Trade-Offs and Contingent Agreements

Even with a good set of options and fair criteria, disagreements about "dividing the pie" will still arise. One way to resolve them is to trade "across" issues that parties value differently. For example, imagine you are involved in a negotiation wherein you could negotiate a trade-off across two issues: a cost-sharing formula and the role of civil society in implementing a treatment program. Imagine that you care more about maximizing the role of civil society, and your counterparts care more about minimizing the cost of implementation. You might therefore accept a higher cost-share, to be borne by you or by civil society. In exchange, your counterpart might accept more civil society involvement in implementation.

Box 3.8 Methods for evaluating options:

- Categorize and prioritize
- Rank order
- Criteria matrix—compare options against criteria
- Highlight advantages and disadvantages
- Ask people, "What do you like about...?"
- Consult decision-makers, community leaders and experts
- Hold a straw vote
- Use "rejection" voting to eliminate less-preferred options

If a direct trade-off across issues is not possible, another way to resolve the disagreement is to make a *contingent agreement*. A contingent agreement is a way for participants who cannot be sure about the impact of an agreement on their interests to reduce the risk involved in the agreement, and put in place a procedure for changing it in response to future developments. For example, imagine that your counterparts are concerned that the civil society stakeholders do not have the capacity to follow through with implementation, and that because the government lacks direct control over their behavior, they might decide to implement the program in a way the government does not like. You and the government officials might agree to try a system that allows the civil society group to take a smaller role in implementation as a pilot project, for the first 6 months. After that trial period, the group as a whole will reconvene to review the progress and agree in advance that if the NGOs implementing the programme meet the agreed-upon criteria, their role would then be expanded. This contingent agreement allows the government to be assured of its interests in quality and oversight, and enables you to expand the role of civil society in the long run.

In order to analyze and make trade-offs that are advantageous for one's side, it is important for negotiators to be very well-prepared regarding their priorities amongst issues. In the Kyoto climate change negotiations, AOSIS anticipated that trade-offs would have to be made and designed its negotiation strategy accordingly. For example, the AOSIS representatives knew that there would not be consensus on their goal of a 20% reduction in emissions by 2005, but by figuring out what they could be flexible about and what *had* to be part of the final package, they were able to maximize the satisfaction of their interests. As a result, the final Kyoto Protocol met many of AOSIS's primary interests, and, most importantly, was legally binding and not voluntary. (The fact that almost every industrialized country was about to fail to meet its voluntary Framework Convention on Climate Change target of 1990 levels by 2000 had convinced AOSIS that voluntary commitments could not be counted on). In addition, while the final Kyoto target was lower than AOSIS sought, it might have been considerably weaker without AOSIS's advocacy of a much stronger target (see Chap. 9).

A *single-text approach* may be useful at this stage to manage the process of making trade-offs and moving the negotiation towards agreement, especially when dealing with a complex set of issues that will require organizational commitments and possibly legal, regulatory and/or policy change. The single-text approach involves centering deliberations on a single, jointly developed draft document, rather than discussing several draft texts at the same time. One strategy that is often effective is to ask countries that have no strong interests at stake in the negotiation, or whose interests are well-represented by other parties, to lead in drafting, allowing them to take on a quasi-mediation role. The single text may outline multiple options for each issue under discussion, placing provisions that are not yet agreed in brackets to be further discussed by the parties. The single text is critiqued—not accepted or rejected, either in part or whole—by the parties and then revised iteratively based on discussions until the draft cannot be improved further. In other words, parties make no commitments until a full package, including trade-offs, is developed. The result is a unified framework document that reflects shared understandings and agreements within the group. In addition, the side-by-side presentation of multiple options in the document allows parties to consider many issues simultaneously. This facilitates trade-offs and encourages the creative mixing and matching of options within and across issues. By compiling agreed-upon and unresolved issues into a single text in this way, the parties can more effectively monitor their progress and avoid competing proposals.

As in many multilateral negotiations, the single-text approach was used in the FCTC negotiating process, where a major role of the chair of the Intergovernmental Negotiating Body was to prepare, issue and revise a "chair's text." In between sessions of the Intergovernmental Negotiating Body, the chair held consultations with delegations, both individually and in groups, to explore interests and possible options. This evolving document incorporated the various proposals for a convention into a single text that served as the basis for the negotiations. Towards the conclusion of the negotiation, as consensus emerged on key issues, a limited amount of bracketed text remained, and the friends of the chair (a small number of key stakeholders concerned about the issues in brackets) got together to make final trade-offs across issues to create a final text on which agreement could be reached.

3.2.3.2 Reaching Final Agreement: Modes of Decision-Making

A number of decision-making processes can be used to move to a final agreement, especially in multi-party negotiations. A successful negotiation is concluded by an agreement among participants, whether unanimous among all participants, or among a substantial number of participants capable of implementing what they have agreed. Unanimity may be desirable but is not always essential for success in a negotiation process. Moreover, while parties may strive for unanimity, it is risky to make unanimity the decision rule. Unanimity rules encourage "hold-outs." With a unanimity rule, one or two stakeholders who are dissatisfied with a tentative agreement are able to block it and demand large concessions as the condition for their support. Instead, it may be better to allow parties to have recourse to some form of voting, such as support by a qualified majority (e.g., two-thirds) of participants. The World Health Assembly, the governing organ of the WHO, for example, has the authority under Article 19 of the WHO Constitution to adopt conventions or agreements within the competence of the WHO by a two-thirds vote.

Alternatively, consensus rules that do not require unanimity, such as "sufficient consensus,"[10] and decision rules that allow stakeholders to opt out or abstain in

---

[10] "Sufficient consensus" was used as a decision rule in the South African constitutional negotiations to end apartheid, in order to ensure that progress could be made amongst multiple parties with differing interests. The consensus required did not involve unanimity, nor an arithmetic counting of votes. Instead, it provided for a consensus that allowed the process to go on to the next stage and did not result in the breakdown of talks—effectively encouraging parties

order to not block progress, are advisable. In these cases, those who are dissatisfied with a tentative agreement should be allowed to state the reasons for their dissent, and other stakeholders should seek to find creative ways to meet the concerns of the dissenters. If solutions cannot ultimately be found that satisfy all stakeholders, those who cannot agree should have an opportunity to record their outstanding concerns, and other stakeholders may want to state the reasons why those concerns were not met in the final agreement. An innovative "opt out" procedure was used in the International Health Regulations to address potential obstacles to global action associated with the need for ratification of treaty obligations within signatory states. While many treaties require ratification at home—in effect "opting in"—the IHR entered into force at the time of their adoption by the World Health Assembly; WHO member states could "opt out" within a specific period of time if they later did not agree; rather than delay adoption of the agreement, the IHR allowed countries to opt out of certain provisions or submit reservations at a later date (World Health Organization 2005, Article 59).

Having cautioned about the risks of a unanimity rule, it is important to note that, when achieved, unanimity can send a very powerful signal and strengthen enforcement. For example, the significance of the FCTC has been enhanced by the unanimous adoption of the final text in 2003, following the likewise unanimous adoption of the initial resolution by the WHA in 1999.

In cases where the deliberating body is providing recommendations rather than making decisions, a final report might distinguish recommendations by their level of support (e.g., full consensus, super-majority and majority support). Alternatively, the issues in dispute can be referred to an independent individual or group that is regarded as competent and legitimate by all participants, and a nonbinding recommendation or binding decision sought on how to resolve the issues in dispute.[11]

3.2.3.3 The Importance of Relationships of Trust

All negotiation processes create relationships. Past interactions influence the way a negotiation is conducted in the present. If one party has handled a past disagreement poorly and their negotiation partners feel they have been unduly pressured, it will be more difficult to establish a good working relationship, and the parties will experience difficulty reaching agreement in a current negotiation. In addition, stakeholders may fear that their goodwill offered during a negotiation could be used against them in the future.

Building (or rebuilding) trust is critical to success in most negotiations. Exploring interests, generating options and distributing value are all made easier

---

(Footnote 10 continued)

to withhold vetoes unless they felt so strongly about an issue that they would leave the negotiations (Mnookin 2003).

[11] See also an example of decision-making by consensus within the Organization for Security and Cooperation in Europe (OSCE) (Chigas 1996, p. 33).

by trust between the parties. Trust is also important when preserving the relationship is as crucial as meeting one's needs in a particular negotiation.

In the simplest terms, the best way to build trust is by demonstrating trustworthiness. For example, if a negotiator states that her government will increase the resources devoted to health surveillance as part of a short-term action plan, it is critical that the resources are in fact committed, and evidence brought to the negotiation table, so that the other negotiators can see that the commitment has been honored. Maintaining consistency between words and actions is one of the most important ways to build trust.

Box 3.9 In dealing with the USTR over pharmaceutical pricing, the Brazilian government adopted a multi-sectoral approach, consulting and involving leaders from a range of sectors within the country to define its policies and negotiation goals. The consultative process helped the government representatives to build trust with the leaders of key constituencies in Brazilian society, and to explore ways to align interests. Through consultation and trust-building, Brazil's negotiators were able to minimize tension between the key constituencies and the government negotiating team during the process (see Chap. 8).

3.2.4 Inclusion of Outside Stakeholders in the Process

Increasingly, the inter-sectoral nature of global public health issues means that a large number of stakeholders need to be at the negotiating table. Input from civil society groups, citizens and beneficiaries, and even industry, is often critical for ensuring that agreements reached are implementable. However, not all stakeholders need to have a place at the table. Stakeholder input may be gathered in a number of ways, including through public hearings, surveys, deliberative polling and citizen juries (Goodin and Dryzek 2006).

The FCTC process shows how a broad group of stakeholders can be effectively included in a transparent way. It openly aimed at encouraging the participation of actors who had been traditionally excluded from state-centric UN governance. In October 2000, the WHO held the first-ever, two-day public hearings that allowed interested groups to register their views prior to the intergovernmental negotiations. This process generated more than 500 written submissions and testimony by 144 organizations, ranging from transnational and state tobacco companies and producers to public health agencies, women's groups and academic institutions. The NGOs served a crucial educative function by organizing seminars and briefings for delegates on technical aspects pertinent to the proposed convention. They also engaged in extensive lobbying activities involving policy discussions with governments, letter-writing to delegates and heads of state, advocacy campaigns, press conferences before, during and after the meetings, and the publication of reports about tobacco industry practices and collusion in smuggling. By acting as the "public health conscience" during proceedings, NGOs became effective advocates of tobacco control. They exposed the obstructionist and dangerous positions certain states took. For example, some organizations called for the U.S. to withdraw from the negotiations given the role of the U.S. delegation in obstructing tobacco control efforts. Some prominent tobacco control advocates were able to promote the public health agenda in the FCTC negotiations themselves as members of national delegations. Public health NGOs and advocates could thus successfully constitute a counterweight to pressures on national delegations by the tobacco industry (Collin et al. 2002). Similarly, during the IGWG, there were public hearings, and WHO Secretariat held a global web-based public hearing to obtain public input on the first draft from as wide a group of stakeholders as possible. Member states, national institutions, health profession organizations, NGO, academic institutions, and others submitted more than 90 contributions to the public hearings.[12]

3.2.5 Culture and Negotiation

Finally, it is important to note that global public health negotiations may also feature cross-cultural communication. Differences in languages, background and cultural norms for negotiation increase the risk of miscommunication. Norms for cross-cultural communication differ greatly as well.

Negotiators from some cultures tend to emphasize explicit communication and formal-process (so-called "low context" negotiation), while negotiators from other cultures tend to communicate more indirectly by using context and informal settings to develop understanding and agreement ("high context" negotiation). Either of these approaches may work well to achieve joint gains when used by all parties. Reaching agreement may be significantly more difficult with both "low context" and "high context" stakeholders at the table, if the negotiators' lack of awareness of the differences leads to misunderstandings and tensions.

In the face of these challenges, some negotiators resort to a "cultural sensitivity" approach that focuses primarily on mastering cultural symbols and styles. They spend significant time preparing for a negotiation by learning the most important cultural norms and signals in order to ensure that their negotiating partners will not be offended or misunderstood. While it is important to note the cultural sensitivities of your negotiating partners, the cultural sensitivity approach

---

[12] *See* Contributions to the First Public Hearing, http://www.who.int/phi/public_hearings/first/en/index.html (accessed 2 December 2010); Contributions to the Second Public Hearing-Section 1, Draft Global Strategy and Plan of Action, http://www.who.int/phi/public_hearings/second/contributions_section1/en/index.html (accessed 2 December 2010); Contributions to the Second Public Hearing-Section 2, Proposals in Response to WHA 60.30, http://www.who.int/phi/public_hearings/second/contributions_section2/en/index.html (accessed 2 December 2010).

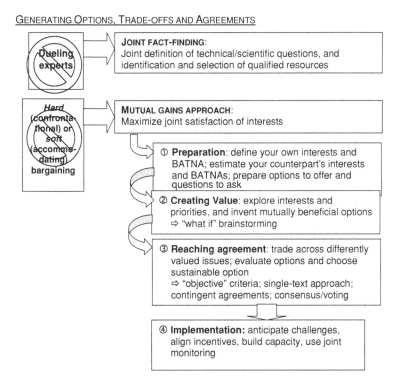

Fig. 3.3 Generating options, trade-offs and agreements

is not a negotiation strategy by itself—cultural awareness must not replace examination of the substance of the issues, the stakeholders' interests and the search for mutually acceptable outcomes.

3.3 Conclusion

In global public health negotiations, the stakes are high for virtually all the parties, and the consequences of a failure to agree are in most cases too grave to be tolerable. Therefore, it is rarely acceptable to delay resolving the issues at hand and/or produce faulty outcomes by resorting to hard bargaining based on a different and irreconcilable reading of the facts. Joint fact-finding processes that build a firm foundation, followed by interest-oriented negotiation using the mutual gains approach and including input from all relevant stakeholders, offer the best chance of generating effective and sustainable health policies. Figure 3.3 summarizes these key negotiation elements.

Chapter 4
Coalition-Building and Process Strategies

Abstract Global public health negotiations are complex, multi-stakeholder processes that cover a wide range of issues. Effective health policies require the consideration of health-related issues in trade, environment and other areas, and are best achieved by involving a broad range of both governmental and nongovernmental actors in the policy-making and negotiation processes (Blouin 2007). Any single country—even a relatively wealthy one—has limited ability to influence such negotiations, or even to make itself heard. A country that joins together with other nations and nongovernment actors, however, can significantly increase its leverage in the negotiation process. Indeed, coalition-building strategies offer effective, and often necessary, means to advance one's health agenda.

Keywords Coalitions · Coalition-building · Public health negotiations · Multi-stakeholder processes · World Trade Organization (WTO) · Doha summit · TRIPS · AOSIS · Framework Convention on Tobacco Control (FCTC) · Framework Convention Alliance (FCA) · Non-governmental organizations (NGO) · Brazil · HIV/AIDS · Coalition membership · Negotiation strategy · Stakeholder mapping tool · Interests · Sequencing · Backward mapping · Spoilers · Blocking coalitions · Clean development mechanism (CDM) · BATNA

4.1 Coalitions as a Source of Negotiating Power

A *coalition* is a group that collaborates to advance the shared and complementary interests of its members. However, coalition members may not agree on all issues, and in fact it is normal for coalitions to engage in ongoing internal negotiations while seeking to maintain a united stance in a larger negotiation process.

International health coalitions and alliances are a central feature of global health negotiations. Coalitions are crucial for promoting health issues in multilateral forums, where multiple stakeholders and multiple issues must be addressed. coalitions can reduce the complexity of those forums by turning a large negotiation with many parties into a smaller negotiation where coalition leaders represent the interests of their members, and prioritize the issues. Coalitions can also help increase the influence of individual countries with limited resources, both by ensuring that their interests are represented, and by demonstrating that their interests are broadly shared.

In this way coalitions can compensate, at least in part, for a lack of structural power. In the negotiation process, coalitions can also enhance members' access to information, pool resources to secure expertise, and coordinate members to ensure that the coalition is represented in working groups, informal meetings and related conferences that are important in a complex international negotiation. Indeed, the breakthrough in negotiations over TRIPS during the Doha summit of the WTO would not have been possible without the leadership and cohesion of the Africa Group, a coalition of more than 50 developing countries that sought to ensure that WTO obligations would not undermine public health campaigns.

Coalition strategies are particularly important for developing countries—and, to an even greater degree, for developing countries' health ministries, which tend to be less influential than finance, trade, or foreign ministries in their own countries. Developing countries in general, and their health ministries in particular, often lack negotiating leverage; few have the resources of their developed country counterparts, and few can field large negotiating teams with deep expertise.

Box 4.1 Coalition have been critical to developing countries' success in ensuring that their concerns are addressed in multilateral forums. As discussed previously, a number of small, relatively powerless island countries formed the Alliance of Small Islands States in 1990 to negotiate climate change more effectively. Individually, those countries would likely not have been listened to at the negotiating table. As a bloc, however, they were regularly expected to comment on each step of the negotiations and thereby became a potent political force in the negotiations (see Chap. 9). In the Uruguay Round of multilateral trade negotiations, the Cairns group of agricultural exporters had similarly enhanced their negotiating power by pooling information and technical capacities, recruiting key agricultural states to their coalition and instituting a systematic structure for cooperation. And in the current Doha round of WTO negotiations, the G20, a coalition of developing-country agricultural exporters led by Brazil, has dramatically reframed the negotiating agenda to focus on phasing out EU and US agricultural subsidies.

Coalitions can likewise increase nongovernmental actors' influence in negotiations. NGOs can gain significant leverage in a negotiating process through their ability to coordinate advocacy—not only with each other, but also with governments participating in formal state-to-state negotiations as the rules of participation evolve to allow greater direct participation by NGOs in formal negotiation processes. A key feature of the FCTC negotiations, for example, was broad NGO participation. The representation of civil society actors in the FCTC was greatly enhanced by the formation and development of the Framework Convention Alliance (FCA). In two working group meetings preceding the formal negotiations within the Intergovernmental Negotiating Body, NGO participation had been mostly confined to NGOs from high-income countries and NGOs representing international health interests broadly. As a response, the FCA was formed as a loose alliance of non-governmental organizations to support FCTC development and ratification. This alliance both improved communication between already engaged NGOs and systematically reached out to new and small NGOs, especially those from developing countries. By February 2003, the FCA had established itself as an important lobbying alliance comprised of more than 180 NGOs from more than 70 countries (Collin 2004).[1]

4.2 Coalitions with Whom?

Given the growing number and influence of governmental and non-governmental, health and non-health actors in negotiations that affect health, coalition-building strategies are more and more complex. Health professionals must negotiate and build coalitions with an increasingly diverse set of actors to achieve their health goals, both "across" sectoral boundaries and "out" beyond government institutions. These actors may include local and international trade, environment, security and other officials, local and international NGOs, multinational companies, academics and scientists, professional organizations, pharmaceutical manufacturers, health and development advocacy and service organizations, and private health management and insurance companies. Non-governmental actors are becoming essential actors in the development and implementation of global health policies—both as coalition partners to promote public health goals and as potential obstacles to achievement of those goals. Indeed, the full potential of non-governmental actors as coalition partners in the health field has yet to be realized. As Garrett (2007, p. 23), notes: "Diseases and health conditions that enjoy a temporary spotlight in rich countries garner the most attention and money. This means that advocacy, the whims of foundations, and the particular concerns of wealthy individuals and governments drive practically the entire global public health effort. Today the top three killers in most poor countries are maternal death around childbirth and pediatric respiratory and intestinal infections leading to death from pulmonary failure or uncontrolled diarrhea. But few women's rights groups put safe pregnancy near

---

[1] The FCA was founded in 1999. As of 2009, it included 350 organizations from more than 100 countries working on the development, ratification, and implementation of the international treaty, the Framework Convention on Tobacco Control (FCTC). *See* http://www.fctc.org/index.php?option=com_content&view=article&id=2&Itemid=9.

the top of their list of priorities, and there is no dysentery lobby or celebrity attention given to coughing babies."

4.3 Developing Countries and NGOs as Beneficiaries of Coalition Strategies

Coalition-building is not only essential for succeeding in "external" negotiations on the international level, but also in the "internal" negotiations that take place within countries. For health policy officials, who often do not have a seat at the international table, coalitions with other ministries in their own country, as well as international agencies, non-governmental advocates, and health agencies in other countries can enhance their influence in their own domestic negotiations on the health impacts of trade, finance, development, environment and other sectoral policies and programs.

Cooperation on the national level with other government agencies can also bridge critical knowledge gaps. For example, Brazil's leadership in global health diplomacy on HIV/AIDS can be attributed to close internal cooperation between the Ministry of Foreign Affairs and the Ministry of Health. Diplomats are not necessarily familiar with specific aspects of drug production and drug pricing policies, so the Ministry of Health's experts filled a significant gap (Kickbusch, Silberschmidt and Buss 2007).

4.4 Building a Coalition

Coalition-building considerations should be part of every negotiator's preparation, and negotiators should be proactive in using coalitions to advance their interests throughout the negotiation process. When considering coalitions, it is helpful to keep in mind a few key characteristics of any coalition: its membership, purpose, level of formality, and level of member commitment. Coalitions can vary widely on each of these characteristics, and the differences can have a significant impact on how effectively a coalition meets its members' interests.

Membership: A coalition can be single-sector (government, NGO, international organization, business), cross- or multi-sector; national, regional or global. Members of single-sector coalitions may find it easier to understand and communicate with each other, while cross- and multi-sector coalitions may benefit from the diversity of their members' perspectives and capacities. Coalitions of national actors often form to advance their country's interests in an international negotiation; regional coalitions may link countries and actors with similar interests in a global issue; and global coalitions may seek to advance a broad group's interests, as the G-77 does in the UN system.

In global public health negotiations, countries from a single region often share primary interests because they face similar health challenges, have similar resource constraints and have bonds of culture and language. Accordingly, regional coalitions are common. In dealing with health issues in non-health forums, health ministry officials and advocates from both developed and developing countries may form coalitions based on shared public health interests and values.

Many coalitions formed to influence international negotiations are led by Foreign Ministry representatives. Often, however, agreements reached will directly affect other Ministries that deal with the issue day to day (such as health ministries, whose access to pharmaceuticals may be affected by the intellectual property provisions of trade agreements), quasi-government agencies, non-governmental and business organizations. Increasingly, coalition membership is not only cross-national among Foreign Ministry representatives, but also cross-sectoral among potentially affected government and non-government actors. Cross-sectoral coalitions are especially important when successful implementation of an agreement will require action by and coordination among many government and non-government actors, not all of whom can participate directly in an intergovernmental negotiation process (see also the discussion on involving implementers in Chap. 5, pp. 82–83). By acting together, multi-sectoral coalitions can have significant influence on inter-governmental negotiations.

Purpose: A coalition may form to influence a single negotiation; influence multiple negotiations; or to provide an ongoing forum for dialogue and promotion of common interests. For example, the world's small island states joined together as a coalition on climate change because they recognized that they all faced a unique threat in the form of climate-driven sea-level rise, which could render them uninhabitable. Most of these countries were part of the British Commonwealth, but the president of the Maldives, Maumoon Abul Gayoom, realized that Commonwealth countries had very diverse interests on the climate change issue making that potential coalition less useful, and that only by banding together could small island states advance their primary concern: to stop climate-induced sea-level rise (see Chap. 9).

Coalitions focused on achieving near-term results often generate higher levels of commitment and investment from their members (as did the coalition that put public health impacts on the agenda for the Doha TRIPS negotiations); coalitions focused on long-term relationship building, (such as the G-77 or the G-7 group of industrial democracies) may build social capital among their members that can be used on many issues, though with less cohesiveness on any one issue.

Level of formality in membership and decision making: A coalition can be highly informal and fluid; structured in membership but without formal rules for decision making; or structured in both membership and decision making. In major multilateral negotiations on economic or environmental issues, some issue-focused coalitions may form and dissolve rapidly (as has happened with commodity-specific negotiations in the WTO), while others become increasingly formal over time, as for example the Alliance of Small Island States has become in the climate change negotiations.

Level of members' commitment to the coalition: Many coalitions are non-binding, with limited ability to coordinate member actions. However, coalitions that seek to maximize their members' collective influence in an international negotiation may seek more explicit commitments from their members to abide by coalition decisions and support coalition positions.

In practice, the membership, purpose, formality and commitments of a coalition and its members may evolve over time. In developing the FCTC, for example, delegates from the WHO's Africa region were the first to participate in the negotiations as a regional bloc. They avoided potential divisions between the tobacco producers and non-producers among them by developing common positions at preparatory meetings prior to each session of the Intergovernmental Negotiating Body. Those coordinated positions combined a broad commitment to tobacco control with recommended measures to assist with producer diversification, and they proved successful in heightening the coalition's impact on the negotiations. This practice was subsequently adopted by other regions, which in turn allowed for developing cross-regional alliances, especially between Africa and Southeast Asia (Collin 2004).

4.5 Coalitions as a Central Element of Negotiation Strategy

An individual negotiator who is considering joining a coalition should consider how well its membership, purpose, rules, level of member commitment and level of influence fit with the negotiator's own identity, interests and resources. More specifically, the negotiator should ask several core questions about any coalition before making commitments to it:

- How well will the coalition be able to advance my own primary interests?
- What trade-offs against my own interests will be necessary to maintain the unity and effectiveness of the coalition?
- What level of investment (time, resources) will I/we need to make in the coalition in order to gain benefits from participating?
- What additional costs/benefits might participation in the coalition have beyond its impact on the negotiation at hand? Other benefits might include relationship building for joint action on other/future issues. Costs might include damage to relationships with others who are outside and opposed to the coalition, or in tension with some of its members.

This chapter explores these questions in more detail.

4.6 Assessing the Added Value of a Coalition

The development of any coalition strategy should begin with an assessment of the potential added value of teaming up with others: of whether it will allow the negotiator to work more effectively and reduce complexity, without raising new and costlier barriers to agreement. Questions to consider include:

- If a coalition already exists, how close are its stated goals and strategies to the negotiator's own? If it is in formation, how much will the negotiator be able to influence the definition of its goals and strategies?
- How do other stakeholders view the coalition? The negotiator should consider both those who are included and those who are outside (whether neutral or opposed to its goals).
- How effective is the coalition in the broader political context of the negotiation?
- What actors would be critically important to bring into the coalition in order for it to have greater influence on the negotiations? What prospect is there of getting those actors to join?
- How much voice and influence would the negotiator have in the coalition? Are key allies in leadership roles, or could the negotiator assume a leadership role?

It is important to identify any potential risks of coalition strategies as well. While coalitions can significantly enhance the leverage of developing countries, coalition strategies can also backfire. Coalitions that lock members into positions can bring a negotiation to stalemate, even where satisfactory options might be available. In addition, if coalitions are perceived to be difficult, developed countries might choose to shun multilateral forums, where developing countries can forge alliances, in favor of bilateral negotiations in which they may have greater economic and political power (as the United States has done in the trade sector, negotiating bilateral agreements with individual countries outside the context of the WTO).

4.7 Mapping Key Stakeholders to Identify Potential Coalitions

To aid in deciding whether or not to join or form a coalition, it is very helpful to have a systematic way of mapping key parties, issues and interests. The following "stakeholder mapping tool" (Fig. 4.1) may be useful in this regard. Using this tool, the negotiator can do a quick initial assessment of the interests and influence of other stakeholders with regard to the goals that the negotiator is trying to achieve. Based on this analysis, he or she can identify potential coalition partners who share interests, as well as potential blocking coalitions that might need to be addressed, and other stakeholders who might be mobilized in support or opposition.

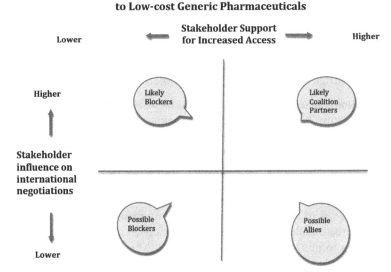

Fig. 4.1 Stakeholder mapping tool for coalition identification

As the figure shows, stakeholders can be grouped into four basic categories: those with high support for the negotiator's interests and high ability to influence the outcome of negotiations; those with low support (or opposition) and high influence; those with low support and low influence; and those with high support and low influence. Conceptually, those with high support and high influence are the negotiator's top priority allies and coalition partners. Those with low support and high influence are potentially members of a blocking coalition, unless some combination of education, advocacy and/or negotiation can shift them to support. Those with low support and low influence may be mobilized to support a blocking coalition, may be persuaded to support the negotiator's goals, or may be ignored. Those with high support and low influence may be mobilized to support the negotiator's coalition.

To take a concrete example, imagine a hypothetical developing country negotiator, representing a health ministry in the WTO TRIPS negotiations on access to pharmaceuticals. The negotiator is seeking to increase access to those pharmaceuticals, perhaps by relaxing current restrictions on compulsory licensing or by seeking voluntary commitments from major pharmaceutical companies to match the price of generic drugs in developing countries. Conceptually, Fig. 4.1 shows how one might begin to draw a stakeholder map of the negotiation process.

To develop the analysis, the negotiator would want to identify major stakeholders in the negotiation process, assess their levels of support for increasing access to generics, and assess their levels of influence on the outcome of the

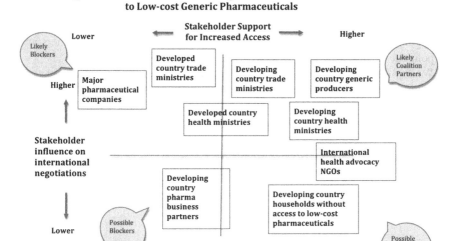

Fig. 4.2 Hypothetical stakeholder map for building a coalition on increasing access to low-cost generic pharmaceuticals

negotiation process. Figure 4.2 shows an initial, rough map of such a hypothetical assessment by the developing country negotiator.

This map suggests some possible strategies for the negotiator representing a developing country health ministry. The negotiator's most likely coalition partners would be other developing country health ministries, producers of generic drugs, and international NGOs focused on improving public health in developing countries.

Now assume that the negotiator has formed a core coalition with other developing country health ministry representatives, and they have jointly developed a stakeholder map like the one above. At this point, they might jointly seek to answer several critical questions about other stakeholders:

- How might we gain the support and commitment of trade negotiators from our own countries? What interests of theirs could we link to our own, so that they see great value in joining the coalition to increase access to generics, and will not trade off that goal in order to achieve other trade and economic development goals in the WTO?

- How might we enlist our colleagues in developed country health ministries, so that they put the global public health interest ahead of narrower national economic interests, and ahead of their existing relationships with pharmaceutical companies in their own countries?

- How might we best partner with international health NGOs to increase our influence on developed country trade representatives and the major pharmaceutical companies? How well-aligned are our goals, and can we use that alignment to create a strategy for influencing public opinion, consumers and political leaders in developed countries?
- What would be the best approach to the major pharmaceutical companies? Should our strategy be to build pressure on them before seeking to engage them, or should we approach them soon to find out the conditions under which they might be willing to collaborate actively in reducing the prices of their medications?

Answering these questions will help developing country negotiators form a coalition-building strategy, one that has the maximum chance of increasing their influence at minimum risk to their top priority interests.

4.8 Developing a Coalition-Building Strategy

Developing a coalition is both a sequencing task and a negotiation task. Once they have mapped the parties, based on available information and subject to refinement as they gain new information, negotiators need to develop a coalition strategy to determine which potential coalition partners to approach and in what sequence. For this purpose, they need to identify not only those parties whose interests are closest to their own, but also powerful and/or influential key players whose involvement and additional leverage is critical to achieve a certain outcome. Starting with the desired outcome, the negotiator identifies key players whose actions and/or decisions are critical to achieving that outcome. The idea is to achieve a critical mass of stakeholders—a "winning" coalition—who together can decisively influence the outcome of the negotiation.

Getting the order and sequencing of coalition-building right is not always easy. Rules of thumb such as "approach allies first" or "gain consensus internally before negotiating with external parties" can be useful, but are not always the best. Backward mapping can be a useful analytical tool to help develop an effective coalition-building strategy (see Fig. 4.3). Backward mapping starts with the end-point and works back to the present to develop a critical path.[2] The idea is to start by asking, "What decisions do we seek?" and then "Who needs to take those decisions or is otherwise critical to whether the decisions are taken?" If the negotiator has limited access to or influence with the decision makers, then one identifies which stakeholder does have the most direct communication and influence with those decision makers and actors, and develops a strategy to attract that "influential" to the coalition. This process continues until a pathway has been

---

[2] See Lax and Sebenius 2006 for a fuller description of "backward mapping.".

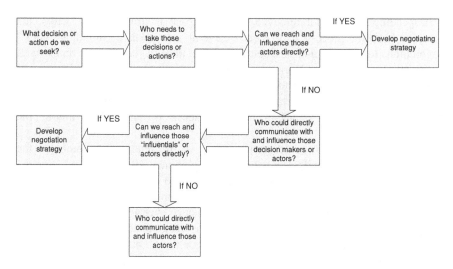

Fig. 4.3 Backward mapping process

mapped from those who can be influenced directly to those who are the ultimate decision makers.

In the example above of the developing country health ministry negotiating access to low-cost pharmaceuticals in the WTO, it may be critical to obtain the support of developed country trade ministries, as they may block any action if they become too closely aligned with TRIPS. At the outset, an individual negotiator may have relationships with only one or two of a much larger group of actors whose support will ultimately be necessary, and no access to or influence with developed country trade ministries. The first step may thus be to form a core group of actors with closely aligned interests. If this core group is composed mainly of less-influential actors, the challenge of bringing in more influential actors remains. The core group of like-minded but relatively low-power core group would thus need to ask a number of questions:

- Who could directly communicate with and influence developed country trade ministries? Developing country trade ministries? Developed country health ministries?
- How can we best use our existing relationships to reach out to those more influential actors whose interests are at least partly aligned with ours?
- Who within our core group would be well-positioned to engage with each of these more influential actors?
- If there is no one within the core group now who has relationships with one of those actors, are there others who do have those relationships whom we could bring into the core group?

Depending on how far removed the core group is from the key actors, the coalition-building process may require several intermediate "building block" negotiations to bring in more-influential players, until the coalition is finally well positioned to approach and engage with the most influential players. The critical challenge in these intermediate steps is to bring in additional allies without compromising the more important interests of the core group in the coalition. This process often requires rounds of internal discussion and analysis among the expanding core group.

Including in the stakeholder map a sense of the relationships amongst the different stakeholders will facilitate the development of a coalition strategy through backward mapping. Who, if anyone, defers to whom? One party may have authority over another, or may have leverage over another, so that the latter will almost certainly do what they do. Who has influence over whom? Who is likely to follow whose lead? And which relationships are antagonistic, such that a coalition or alliance with one party will preclude agreement with another? Understanding these patterns of deference, influence or antagonism amongst the stakeholders can help negotiators determine the best order in which to approach potential coalition partners.

For strategic purposes, negotiators must also decide what information they need to obtain, what to disclose, and how to frame issues and use persuasion skills and techniques to convince other parties to join their coalition. These decisions should be driven by the mapping analysis suggested above, so that each conversation with a potential coalition partner comes after an analysis of whether and how that potential partner might contribute to the coalition and what key questions should be jointly explored.

4.9 Dealing with "Spoilers" and Blocking Coalitions

As a negotiation unfolds, coalition members may realize that other coalitions have formed that have interests in conflict with their own. Also, it may become clear that some stakeholders might not need, or even want, an international agreement of any kind. The role of the tobacco industry in the FCTC negotiations and their efforts to prevent any meaningful tobacco control measures is a case in point.

There are many types of adversaries. They may be parties at the negotiating table, such as the tobacco-producing countries within the FCTC negotiations, or non-participants in the negotiations themselves who attack the process or the results from the outside. These stakeholders may be irreconcilably opposed to an agreement or to the interests the coalition members are pursuing, or they may be "limited spoilers" who have limited goals that they would like to see addressed in the negotiations, and who, therefore, might be accommodated (Stedman 1997). "Greedy spoilers" (Stedman 1997) are in between—they may take advantage of a situation to expand their goals if they see low cost and low risk in doing so, but will limit their goals if the costs and risks become high (Stedman 1997). The Pharmaceutical Manufacturer's Association (PhRMA) could be considered a "greedy spoiler" in the negotiations with Brazil over Brazil's patent laws, as they reduced their demands to ensure their primary interests in safeguarding the principle of protection of intellectual property were met when it became clear that US Government's support would not be unconditional (see Chap. 8).

It is important to identify these types of adversaries early on, and to devise strategies to engage, deflect or isolate them. Options for dealing with blocking coalitions are more complex versions of the same negotiation strategies that can be used in a two-party negotiation, creating agreements that build on shared or complementary interests; using issue linkage to make trade-offs on opposing interests; and reducing the opposing coalition's ability to get what it wants unilaterally (in other words, weakening its BATNA). Each of these options has potential benefits and risks.

Negotiating on the basis of shared or complementary interests: in some cases, stakeholders in the potential blocking coalition may not have considered fully the potential for mutually beneficial outcomes from the negotiation process; or such outcomes may emerge through an initially adversarial negotiation process. For example, Brazil's decision to build its capacity to manufacture anti-retroviral (ARV) drugs and to invoke its rights under the WTO compulsory licensing clause, triggered an adversarial series of negotiations and dispute resolution actions with the US Trade Representative and the US pharmaceutical industry. Though some aspects of the negotiation process remained adversarial, several joint gains did emerge. Most notably, a series of informal price discount agreements between pharmaceutical companies and Brazil (and eventually other countries as well), set a precedent that evolved into a norm of differential pricing.

Differential pricing significantly reduced the number of trade disputes between the US pharmaceutical industry and developing country governments; gave the pharmaceutical industry direct control over pricing decisions (something it forfeited when developing countries exercised their compulsory licensing option); and gave developing countries access to ARVs with lower transaction costs and relationship costs than would have been the case if they had to use the WTO dispute resolution process (see Chap. 8).

Differential pricing was not an unambiguous gain for either developing countries or the US pharmaceutical industry. Some developing countries may have foregone opportunities to get even lower prices by licensing domestic manufacture. Some pharmaceutical companies may have seen their profitability cut more than it would have been if the USTR had won their cases through the WTO dispute settlement mechanism. But there were enough shared interests in reducing the cost of dispute resolution, and complementary interests in certainty of access and certainty of pricing, to create a joint gain through agreements on differential pricing.

Issue linkage as a way to make trade-offs: Sometimes interests are sharply enough opposed that blocking coalitions have to be compensated for making substantial trade-offs on issues that are very important to them. A very literal example of this strategy is the agreement under the Kyoto Protocol to create a Clean Development Mechanism (CDM) for funding carbon emissions reductions in developing countries. Developing countries advocated forcefully in the Kyoto Protocol negotiations for developed countries to commit to substantial emissions reductions, to be achieved by changes in their domestic regulations and incentives for energy use. They refused to take on any emissions reduction commitments of their own, on the argument that the developed countries had created the climate problem. Developed countries argued that developing countries also needed to take action, as their emissions were increasing at a faster rate, and in total would soon overtake those of the developed countries.

The CDM emerged as a way for developed countries to make good on their own emissions reduction targets, by paying developing countries for undertaking projects that would reduce their emissions, and counting the reductions against the developed countries' own emissions reduction targets. In this way, the CDM served as a very direct compensation mechanism to resolve a difficult trade-off between developed countries who wanted to reduce emissions at the lowest possible cost, and developing countries who wanted the developed countries to take full responsibility for emissions reductions. Paying for projects in developing countries through the CDM helped resolve what might otherwise have been an unbridgeable divide.

Weakening a blocking coalition's BATNA: In general terms, the idea of weakening a negotiating partner's BATNA is the same with coalitions as it is for individual negotiators. However, when two coalitions are negotiating with each other, it is possible to weaken a potential blocking coalition by splitting its members. When a coalition loses members, it generally loses some of its influence and its ability to impose unilateral solutions. Coalitions can be split by "side-deals" with individual members or sub-groups, or by putting coordinated pressure on them. For example, developed countries have used bilateral deals and incentives to split coalitions of developing countries on numerous issues. Some developing countries in the ARV licensing negotiations (referred to above) were persuaded to leave the coalition headed by Brazil, based on offers of low-cost pricing from pharmaceutical companies. Other negotiators decided that the risk of negative consequences in their broader trade and aid relationships with developed countries was too high, and left the coalition as a result (see Chap. 8).

4.10 Multi-Stakeholder Coalition Building

With the increasing importance of nongovernmental actors in both the domestic and international arenas, coalition-building (and blocking) processes have become more complex. At the same time, coalitions bringing together government, international organizations and NGOs have also opened tremendous opportunities for health actors to enhance their leverage.

At one level, nongovernmental actors and narrowly focused international organizations may complicate the process of defining interests and priorities and pursuing goals. If single-issue organizations have agendas that overlap with but are not identical to developing countries' agendas, they can undermine developing countries' ability to pursue health priorities and economic growth, since these organizations often do not have to balance multiple interests. For instance, the 60-member coalition of developing countries pushing for exceptions to TRIPS for public health priorities risked fracturing because of external pressures from NGOs that were focused on the single issue of access to essential medicines.

Yet NGOs and developing countries can work together effectively to advance shared and complementary interests. Jointly, they can frame issues from both a fairness and a public health perspective. NGOs can mobilize public opinion and put sophisticated media and consumer pressure on potential private sector blockers. Developing country representatives can build on effective issue framing by advancing proposals on the basis of fairness and compassion, as well as public health needs. Multilateral health organizations can sometimes provide useful technical assistance, as long as they do not overstep their mandates or jeopardize their relations with member states. The larger and more cohesive the coalition of governmental and NGO advocates, the more effective their combined mobilization and negotiation capacity.

For example, developing countries and NGOs together used the media and access to sympathetic politicians in the United States and Europe to shape the terms of the debate on intellectual property rights and public health and to influence the US and the EU to modify, or at least publicly justify, their positions within the WTO (Shadlen 2004). Even more importantly, NGOs and international organizations can provide critical analytic resources and legal and technical expertise and assist in the formulation of collective goals and strategies. Organizations such as the Foundation for International Environmental Law provided this kind of expertise to AOSIS during the Kyoto Protocol negotiations. Activist organizations such as Médecins Sans Frontieres did the same for the developing countries' initiative on intellectual property and public health in the TRIPS Council (Shadlen 2004). Serving as the secretariat for the Framework Convention on Tobacco Control, WHO used both its convening authority and its technical expertise in ways that complemented the efforts of governments and NGOs seeking limits on tobacco marketing.

In the TRIPS case, developing nations urged an approach to the TRIPS agreement that would provide them with a greater degree of flexibility when dealing with matters related to public health, particularly the HIV/AIDS pandemic. Some developed countries, led by the United States, did not want to create health exceptions to intellectual property rules, arguing that protection of patent holder rights was vital to providing incentives for the creation of new drugs. In this setting, a powerful alliance between developing countries and NGOs fundamentally transformed the intellectual property debate within the WTO from one on substance to one about procedure. Reducing developing countries' obligations for patent protection was not feasible given the economic and political weight of the most developed countries in today's global economy. Therefore, this coalition emphasized procedures as a strategy for protecting the space to take advantage of flexibilities granted by TRIPS in order to secure access to affordable medicines. In particular, developing countries demanded clarification of countries' rights to issue compulsory licenses and to authorize parallel importation under TRIPS, to reduce the reluctance to use those rights because of the agreement's ambiguities and fear of litigation and sanctions for violation of the agreement.

NGOs were critical to this effort. Key activist organizations helped developing countries form a common position in the TRIPS Council that articulated how countries could use the built-in flexibilities of the agreement. The coalition succeeded in having WTO members' obligations and rights under TRIPS clarified in the Doha Declaration on the TRIPS Agreement and Public Health. The Declaration ultimately reflected a consensus view that governments should construe the TRIPS agreement in a way that supported the realization of their public health goals.

Besides providing legal and technical expertise to developing countries, the NGOs also brought media attention to the implications of stronger patent protection for the treatment of HIV/AIDS. In fact, the process leading to the Declaration had started with a long awareness campaign by civil society groups on this topic. During the 1990s, NGOs such as Médecins Sans Frontieres, Health Action International and Consumer Project on Technology engaged themselves in lobbying efforts aimed at ensuring lower prices for essential medicines for developing countries in need. These NGOs convened representatives of the pharmaceutical industry, other NGOs, national governments and intergovernmental organizations in Geneva in March 1999. At this conference, the potentialities of compulsory licensing under Article 31 of the TRIPS agreement were discussed. By 2001, there was growing public sentiment that people had a right to medicine and that this right was being violated in numerous instances in developing countries. Drug companies were portrayed as avid profiteers at the expense of human lives, and governments that sought greater protection of intellectual property rights were being increasingly criticized. It was this setting that served as the backdrop for the commencement of formal negotiations in the TRIPS Council in early 2001, and framed the issue (as discussed previously in Chap. 2) in a way that made it easier for the NGO-developing country coalition to reach its goals (Shadlen 2004).

As noted above in the discussion of blocking coalitions, Brazil forged an informal global coalition with HIV/AIDS and human rights NGOs, African countries fighting the HIV/AIDS epidemic, and international organizations to challenge the patent rights of US pharmaceutical companies, ultimately leading to a change in US policy. Brazil's strategy attracted enormous international and media attention for the HIV/AIDS issue and created the perception that the USTR was the lackey of the pharmaceutical industry. This strategy succeeded in driving a wedge between PhRMA, the leading US pharmaceutical trade association, and the USTR, which was subjected to intense pressure from the media and other domestic groups to change the US position (Chigas et al. 2007).

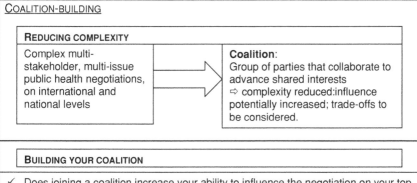

Fig. 4.4 Multi-stakeholder coalition building considerations

4.11   Conclusion

Coalitions can be powerful tools to amplify the influence of individual countries and organizations. They also help negotiators manage the complexity of multi-party, multi-issue negotiations on global public health issues. Strong coalitions can help weak parties to create a level playing field (Fig. 4.4).

A prerequisite of effective coalition-building is thorough analysis via the mapping of parties, issues, interests and influence. A carefully mapped negotiating environment facilitates identifying, building and maintaining winning coalitions; deciding whether to join existing coalitions and assessing whether and when to leave a coalition. In addition, it is important to recognize the importance of sequencing and periodic reassessment in coalition-building. Coalition-building is an inherently dynamic process in which conditions change and new coalitional possibilities constantly arise, so negotiators need to recalibrate coalition strategies in order to advance their health goals most effectively.

Chapter 5
Meeting Implementation Challenges

Abstract The implementation of a negotiated agreement—including follow-up, monitoring and revision as needed—is step four of the mutual gains approach. Crafting an agreement on paper that all parties can support "in principle," without facing the core question of whether and how they are prepared to implement it, generally leads quickly to implementation failures, loss of credibility for the agreement, mutual recriminations among the parties and skepticism about the potential for renegotiation to produce a better outcome. Moreover, many international agreements have "free rider" problems: that is, if most countries comply with an agreement, some countries may be able to violate it while still benefiting from the compliance of the others. For example, countries that have committed to limit greenhouse gas emissions may decide that keeping the costs to their industries low is a higher priority than reducing their emissions—all the while hoping that other countries will continue to hold their industries to their commitments. Negotiators therefore should lay the groundwork for post-negotiation success during the negotiation process. It is crucial to anticipate possible implementation challenges and address them before reaching an agreement. These include unresolved conflicts of interest, insufficient human and financial resources, difficulty monitoring effectiveness, unanticipated new factors and institutional resistance to change.

Keywords Negotiation challenges · Implementation · Agreement · Stakeholders · Contingency provisions · Spoilers · Planning for implementation · Kyoto Protocol · TRIPS · Contingent agreements · Convention on International Trade in Endangered Species of Wild Fauna and Flora (CITES) · Agreement adaptation · AOSIS · Dispute resolution processes · Resource scarcity · Global polio eradication initiative · Global Fund to Fight AIDS, Tuberculosis and Malaria · Malaysia · Institutional resistance · Winning coalition

D. Fairman et al., *Negotiating Public Health in a Globalized World*,
SpringerBriefs in Public Health, DOI: 10.1007/978-94-007-2780-9_5,
© The Author(s) 2012

To help achieve "implementability," negotiators should first ask a set of basic questions about any potential agreement, well before committing to it:

- Does this agreement satisfy all or nearly all of the key stakeholders' primary interests (stated and unstated)? How well? An agreement that is strongly in the interest of all stakeholders to implement will be *nearly self-enforcing*—in other words, does it meet the interests of the main interested stakeholders (whether they are at the table or not) sufficiently so that they have incentives to comply without further enforcement mechanisms?
- Are there mutually acceptable ways to monitor implementation—both to verify that parties are meeting their commitments and to confirm that the actions they are taking are having the desired impact on the problem?
- Does the agreement include contingency provisions that anticipate and address potential implementation challenges or changes in circumstances that may affect the agreement?
- Does the agreement include procedures or mechanisms to assist the parties in answering questions and resolving disputes about implementation?
- Do the implementation stakeholders (not only the representatives around the table but also the agencies and organizations that will be directly responsible for taking action) have the resources, capacities and incentives necessary to take the actions that they would be committed to under this agreement? If not, does the agreement provide a mechanism to mobilize resources, build capacities and/or change incentives?

If the answer to any of these questions is "no," the negotiators should focus on the issue in question and find a way to address the implementation challenge.

5.1 Involving Implementers

In crafting implementation provisions, it is often very important to consult directly with implementing agencies and organizations. One reason is that implementers have their own interests and concerns, which may not be fully represented by high-level diplomatic or ministerial delegates negotiating the agreement. Failure to consider these interests may lead to implementation problems, whether through passive resistance or active rejection by implementers.

Another reason to engage with implementers is to generate more detailed information about potential implementation challenges and opportunities. For example, the effectiveness of an HIV/AIDS treatment might depend heavily on trained nurses' and health aides' ability to help patients learn and maintain the treatment regimen. If the negotiators from the Ministry of Health consulted with representatives of nurses and health aides during their negotiations with a multinational pharmaceutical company for the provision of a complex set of drugs, they might find that the potential treatment regimen is only going to be workable for the target population if the frequency of doses and number of pills can be reduced. That knowledge could (and should) affect the negotiations in significant ways—for example, regarding what drugs are provided, in what form, and what training or support is provided to the MOH.

Finally, "spoilers" and others who were interested in blocking an agreement at the negotiating stage can frustrate implementation by negotiating with the implementers. In the climate change negotiations, for example, following agreement on the Kyoto Protocol, those opposed to the Protocol negotiated with national governments for the issuance of permits for trading of carbon emissions—a permissible action under the agreement. However, because these negotiations with implementers led to the issuance of many more permits than needed (or anticipated), the regime created by the agreement to reduce emissions was significantly undercut. Because so many permits were issued, the price of permits was very low, and incentives to reduce carbon emissions were correspondingly weak (Cooper 2010). Involvement of implementers at earlier stages of the negotiation of the agreement itself might have prevented, or at least mitigated, these "spoilers" ability to block agreement by undermining implementation.

The remainder of this chapter discusses the most common challenges to the successful implementation of public health agreements and offers advice on how to deal with them.

5.2 Insufficient Planning for Implementation

Implementation considerations often become secondary in the heat of negotiations—a mere afterthought detached from the rest of the effort. The result may be an agreement that stands on its own, but without a realistic chance of successful implementation because of a disconnect between the design of the initiative and how it may realistically unfold in practice.

Implementability should be a vital consideration at every stage of the negotiation process. In assessing strategic options, negotiators need to take into account the environment into which each option will be introduced, including the stakeholders, decision-makers and other relevant individuals, institutions and organizations that will need to be involved in carrying out the obligations agreed in the negotiations—both on one's own "side" and on the other's. Careful stakeholder mapping in this regard will effectively reduce the chance of shortcomings in implementer performance. If key stakeholders in implementation who are not at the table have incompatible interests or insufficient incentives, knowledge or capacities to follow through on the negotiators' commitments, implementation will suffer. Negotiators should consider processes to engage implementers in and inform them of ongoing discussions (or include them at the table), as well as provisions to ensure that the capacities and incentives (penalties, rewards) to implement the agreement are also considered and included. It is also important to devise a means of holding the parties to their commitments.

The emissions trading mechanism contained in Article 17 of the Kyoto Protocol illustrates the potential for creative solutions to promote the implementation and sustainability of agreements. It allows parties to acquire emissions units from other parties and use them to achieve compliance with their own emissions targets under the Protocol. Parties can thus take advantage of lower-cost opportunities to reduce emissions in order to decrease overall reduction costs. By recognizing that countries would likely encounter difficulties with domestic stakeholders in reducing emissions within the prescribed time frames and by allowing this alternative path to compliance with the Protocol, the negotiators significantly enhanced implementation and sustainability of the agreement.[1]

For agreements that are not highly compatible with the interests of key implementers (and therefore nearly self-enforcing), legal or regulatory changes or organizational capacity-building may be required for full implementation. In these cases, it is very important that negotiators specify the steps to be taken for implementation and who will take them, to ensure that the agreement will be formalized and implemented. For example, TRIPS recognizes compulsory licensing (i.e., the use of an invention without the permission of the patent holder) as a public health safeguard against a patent holder that charges excessively high prices in a particular market. However, TRIPS is a framework agreement that needs to be operationalized through national laws. Only if incorporated into national law can the compulsory licensing safeguard be used (Reinhardt 2006).

5.3 Common Challenges to Implementation

5.3.1 Failure to Address Uncertainty

Global public health policies are often created and implemented in volatile contexts and settings of uncertainty, in which surprise developments occur that may need to be addressed immediately. A new disease may surface and spread rapidly, as in the case of SARS or the avian flu (H5N1) and swine flu (H1N1) pandemics. Besides changes in circumstances, surprises can include failure on the part of some parties to live up to their commitments, or, more positively, a new opportunity to achieve negotiators' joint goals through a different strategy.

5.3.1.1 Dealing with Uncertainty about the Future: Contingent Agreements

In Chap. 3, contingent agreements were described as a technique for dealing with disagreements and uncertainties that keep parties from reaching agreement. In the

---

[1] For more on this emissions trading mechanism, *see* http://unfccc.int/kyoto_protocol/mechanisms/emissions_trading/items/2731.php.

context of implementation, contingent agreements enable parties to anticipate and plan for potential future changes that can affect implementation. Contingent agreements address uncertain future conditions that may affect the implementation of an agreement. They take the form of: "If condition X, then [a party or parties] take action Y." For example, "If oil prices increase by X%, then the pharmaceutical company will pay Y% of the additional cost to ship the medicine."

Contingent agreements provide several benefits. As noted above, they can help parties overcome divergent views of the future that might otherwise prevent agreement. Agreeing in advance on action to be taken in response to certain predictable circumstances can also help avoid delays in implementation, if not complete breakdown of an agreement, should those circumstances arise. And contingent agreements foster overall implementation success by limiting risk for the parties; they can specify exactly, and limit, parties' obligations under specific circumstances.

The contingent agreements approach has been used to implement the Convention on International Trade in Endangered Species of Wild Fauna and Flora (CITES). This international agreement seeks to ensure that international trade in specimens of wild animals and plants does not threaten their survival. The Agreement takes account of and plans for the likelihood that the threat to the survival of any species will change over time (although it is uncertain how). It plans for this contingency by providing that species enjoy a level of protection under CITES that corresponds to the level of threat they face; if the threat to a species diminishes, its protection level is lowered accordingly.[2]

5.3.1.2 Dealing with Implementation Deficits

Partial implementation or non-implementation of an agreement can result from a number of issues—from lack of political will to lack of capacity to unexpected changes in the environment. A failure to identify implementation problems early on and to adapt can result in partial implementation or non-implementation by a party. Many agreements utilize monitoring and reporting mechanisms for early detection of implementation deficits in the face of unpredictable (or predictable) surprises. For example, the FCTC requires parties to submit periodic implementation reports to the Conference of the Parties for review. These reports must include information on legislative, executive, administrative or other measures taken; constraints or barriers encountered and measures taken to overcome them; and financial and technical assistance provided or received for tobacco control activities.[3] Civil society actors can be valuable partners, as well as sources of data,

---

[2] For more on the CITES agreement, see http://www.cites.org.

[3] See esp. Articles 21 and 23(5) of the FCTC at http://www.who.int/tobacco/framework/en/. *See also* http://www.fctc.org/index.php?option=com_content&view=article&id=131&Itemid=147 (website of the Framework Convention Alliance), for further information about reporting on the FCTC. At its first meeting, the Convention of the Parties established more detailed reporting

in monitoring implementation as well. Implementation of the FCTC benefits from the activities of the Framework Convention Alliance (FCA), an alliance of non-governmental organizations concerned with smoking initially created to facilitate and enhance civil society participation in the FCTC negotiations. The FCA provides technical assistance and capacity building to governments, and has also undertaken monitoring of the implementation of the FCTC as one of its key activities; its status reports on implementation of the agreement in a yearly publication, the FCA FCTC Monitor, is one way of holding governments accountable to their FCTC obligations.[4]

Periodic monitoring and review are essential to assess whether implementation is achieving the group's goals and to respond to new information and circumstances. If the parties have made contingent agreements, monitoring of the conditions that could trigger action is essential to implementation. Ideally, monitoring systems should be joint (i.e., representatives of all key stakeholder groups should be involved), and should periodically seek to assess the extent to which the agreed actions are achieving their underlying goals. In this sense, monitoring can be understood as a continuation of joint fact-finding.

As negotiators design a process to monitor tangible changes, they should include in the agreement indicators of success and means for gathering information on those indicators on a regular basis. For example, if an agreement is meant to achieve a 10% decrease in HIV infections in a given region, it will be important to include a provision in the agreement on how HIV infections in the region will be measured, by whom, how often, and who will be responsible for sharing that information with all the parties. If there is either mistrust between the parties or a lack of internal capacity to undertake the monitoring, negotiators may agree to fund an external party to be the monitor.

5.3.1.3 Adaptation and Revision of the Agreement

An unexpected change in the implementation environment may require revision of the agreement. Negotiators should thus ensure that they have an agreed procedure in place to amend the agreement if necessary. Such a procedure can be a multi-stage effort, including the evaluation of implementation shortcomings and

---

(Footnote 3 continued)
requirements and forms, and created a permanent Convention Secretariat within WHO. Conference of the Parties to the WHO Framework Convention on Tobacco Control, Decisions and Ancillary Documents, First Session, 6-17 February 2006, available at http://apps.who.int/gb/fctc/PDF/cop1/cop1_06_cd_decisionsdocumentsauxiliaires-en.pdf. The Convention Secretariat is expected to provide feedback to each reporting Party, and to provide an annual summary analysis that reflects international and regional progress in implementation of the FCTC, highlights significant achievements, and reflects the spirit of shared learning. The Secretariat's first synthesis of Party reports was submitted to the second session of the COP.

[4] *See* http://www.fctc.org/index.php?option=com_content&view=article&id=136&Itemid=155 on implementation monitoring by the FCA.

determination of their causes, and ultimately reassembly of the parties to amend the agreement.

Periodic meetings of the parties can increase the likelihood that the parties will be able to identify and address implementation challenges. In addition to identifying implementation problems early on, these meetings can foster stronger long-term relationships and reduce the risk of some representatives perceiving others to be unresponsive if difficulties arise.

Making a commitment to periodic review and revision during implementation may also allow parties to agree on an implementation framework that does not fully resolve implementation issues, but lays the groundwork for "learning by doing." Negotiations on the Kyoto Protocol are a good example of "learning by doing". Initially, the implementation mechanisms (such as international emissions trading, joint implementation and the clean development mechanism) were met with reservations from members of Alliance of Small Island States (AOSIS), who faced potentially devastating consequences from a sea level rise of even a few inches (See Chap. 9). "Nobody knew what emissions trading meant," said Chairman Tuiloma Neroni Slade. "By their very description, nobody knew how trading would work and how it would control emissions." Their qualms were not completely quelled by the end of the Kyoto summit because the details had not been worked out in Kyoto but AOSIS negotiators knew they would have an opportunity to shape these mechanisms in future negotiations. Similarly, AOSIS had ensured that a compliance mechanism existed, even if it was also left unfinished. Marshall Islands negotiator Espen Ronneberg became the Co-Chairman of the working group on the compliance mechanism in the first round of negotiations after Kyoto, after demonstrating knowledge and skill in the discussions on compliance during the Kyoto talks (See Chap. 9).

5.3.1.4 Anticipation of Disputes

Both monitoring and revision efforts can lead to disputes. Therefore, regardless of the specificity of the implementation, monitoring and review procedures, negotiators should also include dispute resolution clauses in their agreements. Article 27 of the FCTC, for instance, mandates the diplomatic settlement of disputes on interpretation or application through negotiation, good offices, mediation, conciliation or *ad hoc* arbitration. By agreeing in advance on how to deal with disputes, parties can address conflicts more effectively should they arise.

5.3.2 Resource Scarcity

The most common, or at least the most cited, implementation challenge is a lack of resources for properly carrying out the plans as initially designed. Resource scarcity can cause frustration and rejection of the agreement among those directly responsible for implementation. Over time, inadequate resources will delay or prevent the achievement of results, as well as decrease the agreement's credibility and effectiveness.

In developing an agreement, negotiators should consider the financial costs and technical, human and organizational resources that will be required to follow through on the commitments within it. Negotiators should also assess their partners' authority to commit the necessary resources. Where developing countries are expected to take on substantial implementation responsibilities, they may require resource transfers to build their capacity. For that reason, many North–South economic, environmental and health agreements include mechanisms for North-to-South resource transfers. For example, the UN's Framework Convention on Climate Change established a centralized mechanism to channel financial assistance to developing countries through the Global Environment Facility; the same mechanism is being used to implement the Kyoto Protocol.[5] Similarly, the FCTC Convention Secretariat and the Tobacco Free Initiative (TFI) at WHO are providing technical support to countries to assist them in their efforts to strengthen their infrastructure and take the necessary steps toward the signature, ratification and implementation of the FCTC.[6] The IHR includes obligations for member states to collaborate with each other in the provision of technical cooperation for the development of the public health capacities required by IHR, as well as in the mobilization of financial resources to facilitate implementation of their obligations under the agreement.[7] The IHR also provides explicitly for WHO to provide technical assistance to states in developing the required control and surveillance capacities called for under the agreement.[8]

One effective way to increase implementation resources is to bring in "resource-rich" stakeholders. Advocates for action should begin early in a process (as early as issue framing) to identify and engage stakeholders who might be able to commit resources. By the time negotiators are sitting around a table, they should already have a shared sense of the potential for resource contributions (financial, technical and/or human) among the parties. However, the development of new options during a negotiation process may lead negotiators to seek additional "resource partners" who can help implement an agreement.

In this context, it is often as important for negotiators to create a sense that responsibility for providing resources is being shared equitably and fairly among the parties, as it is to reach agreement on the amount of resources needed. The process of devising a formula for "fair shares" in resource commitments is usually a central and time-consuming aspect of international health negotiations.

---

[5] See http://unfccc.int/cop7/issues/convkpfunding.html.

[6] *See* Convention Secretariat of the WHO FCTC, Implementation Assistance and Partnerships, at http://www.who.int/fctc/secretariat/implementation/en/index.html and World Health Organization, Tobacco Free Initiative, Strengthening of National Capacity for Tobacco Control, http://www.who.int/tobacco/training/en/.

[7] IHR 2005, Article 44.

[8] *Id.*,

Identifying appropriate objective criteria for distributing the burden can also facilitate agreement about resources. In addition, negotiations may need to identify additional stakeholders who can help lighten the load for others.

Box 5.1 Broad-based resource commitments were critical to the Global Polio Eradication Initiative launched by the WHO in 1988. Addressing an urgent stakeholder consultation on polio eradication on February 28, 2007, WHO Director-General Margaret Chan argued: "As an international community, we have few opportunities to do something that is unquestionably good for every country and every child, in perpetuity. Polio eradication is one of these opportunities."[9] In response, Rotary International and the Gates Foundation provided $200 million for the intensified push to eradicate polio, with the publicly expressed hope that their shared commitment would inspire and challenge other donors and polio-affected countries themselves to ensure rapid mobilization of the necessary financial resources (Global Polio Eradication Initiative 2007).

Few global public health challenges involve single players that have the funding, research and delivery capabilities for solving any problem on a world-wide scale. For that reason, multi-stakeholder alliances have been formed to reduce the burdens of AIDS, tuberculosis, malaria, polio, river blindness and many other diseases (Bill & Melinda Gates Foundation 2002). In these alliances, government, business and NGO partners contribute different kinds of resources, and their complementarity may be essential to both creating a sense of equity and implementing the agreement.

The Global Fund to Fight AIDS, Tuberculosis and Malaria, an independent partnership between governments, civil society, the private sector and affected communities and established in 2002 by the United Nations, is a successful example of a global public health initiative supported by multi-stakeholder contributions. As of July 2009, the Global Fund had committed $19.3 billion in 144 countries to support large-scale prevention, treatment and care programs targeted at all three diseases. It does not implement programs directly, but relies instead on the knowledge of local experts and provides them with the resources necessary to implement programs to prevent and treat AIDS, tuberculosis and malaria. The success of the Global Fund is reflected by both the prominent "Product Red" campaign launched in January 2006 and a $500 million contribution by the Gates Foundation in August 2006.[10] Product Red created an innovative financing mechanism for R&D for diseases of the poor through sales of franchises to consumer

---

[9] Http://www.polioeradication.org/fundingbackground.asp.

[10] For more information on The Global Fund, see http://www.theglobalfund.org and Http://en.wikipedia.org/wiki/The_Global_Fund_to_Fight_AIDS,_Tuberculosis_&_Malaria.

groups who design specific red products, with 40% of the profit from the sales of these products going to the global fund for AIDS, TB and Malaria.[11]

The international corporate world can be of great assistance in surmounting the resource barrier. The onchocerciasis (river blindness) treatment program, for example, grew out of the conclusion by Merck & Co, Inc., that the usual marketing of ivermectin (Mectizan) would not get the drug to those in rural Africa who could benefit from this treatment. In 1987, Merck offered to donate the drug if a mechanism could be found for effective distribution. An independent Mectizan Expert Committee was established to review applications and approve the release of the drug with appropriate safeguards. Through this committee, more than 100 million tablets have been donated, over 20 million people received treatment in 1997 alone, and the prevalence of blindness secondary to onchocerciasis is decreasing. Merck used its scientific and economic resources to target an inequity and has provided an important medical and social benefit to a population that would otherwise have been untreated. The initiative by Merck resulted in a coalition of global organizations, ministries of health, foundations, mission groups, community organizations and volunteers, held together by a shared goal rather than a true organizational structure (Foege 1998).

Successful mobilization of resources for such global health alliances requires a clear, specific and compelling goal, such as the reduction of malaria incidence by 50 percent by 2010. Also useful is a clear scope, as defined in terms of geography, patient populations, functional activities and time. In the past, many successful alliances, such as the International Trachoma Initiative and the Mectizan Donation Program, started with a narrow scope and then expanded as success accumulated. Beyond ensuring clear overall goals and a focused scope, managers and donors seeking to maximize resource commitments for global health alliances should be guided by five questions:

1. Is there a clear understanding of the added value that comes from being in an alliance—and what is required to capture this value?
2. Have the partners selected an appropriate alliance structure? Simpler and looser structures are appropriate where the level of integration or coordination is limited; more complex, tighter structures should be used where the potential value is substantial and where a higher degree of coordination or integration is required.
3. Have the partners gone beyond a statement of shared objectives and also agreed on specific success metrics, milestones and partner contributions?
4. Have the architects of the alliance resisted the urge to have equality for all, instead creating governance models that allow input by stakeholders while ensuring effective decision-making?

---

[11] For more information about Red and an equally innovative program begun by WHO, UNITAID (where a small tax on your airline ticket is set aside to fight AIDS, TB and Malaria), see Pilippe Douste-Blazy and Daniel Altman "A Few Dollars at a Time: How to Tap Consumers for Development", Foreign Affairs (January/February 2010).

5. Is the alliance characterized by a sufficient number of operating staff whose primary objective is the alliance's success? Or, by contrast, are several busy people in different organizations sharing the "chief executive officer" role? Are operating staff or secretariat staff contributing to the alliance on a predominantly part-time or even nights-and-weekends basis? (Bill & Melinda Gates Foundation 2002).

5.3.3 Institutional Resistance

Any agreement that requires institutional change to meet a new global public health challenge will produce some resistance from some of those who are being told to change. There are two principal sources of resistance. First, resistance may stem from a lack of institutional capacity—lack of resources, personnel, skills, knowledge, surveillance or data gathering systems, among others. Asking an organization to perform a task that exceeds its current abilities tends to lead to resistance from that organization and the individuals that make up that organization. Strategies to deal proactively with resource scarcity have been presented above. In addition, there may be training or coaching opportunities to break barriers of resistance, and unwilling implementers may be engaged through effective dispute resolution mechanisms.

Second, resistance may be a consequence of a lack of sympathy for the policy goal or means. Policy initiatives may appear to be contrary to the entrenched interests of a particular organization or to certain parts of it. These conflicting interests, whether real or perceived, serve as a disincentive for cooperation. Resistance based on conflicting interests is best prevented and overcome by addressing all stakeholders' interests during the negotiations, including the interests of potential implementing agencies. It is often possible to address the concerns and interests of organizational leaders through a combination of policy commitments (to address inter-ministerial concerns about the impact of new health initiatives on their ministries), leadership incentives (e.g., new titles, promotions, public visibility and/or increases in organizational budget) and performance incentives.

The example of Malaysia illustrates the importance of leadership by ministries of health and their possible key position in trade negotiations in order to ensure policy coherence. In this case, the Malaysian Ministry of Health played a proactive role in the decision to import generic antiretroviral drugs under the "government use" provision of TRIPS. The ministry faced strong opposition even within the national government cabinet, due to concerns that such action would deter future foreign investment in Malaysia. Thanks to strong political support, though, the cabinet was convinced and authorization was obtained to import these drugs for a period of two years, beginning November 1, 2003 (Blouin 2007; Musungu & Oh 2005).

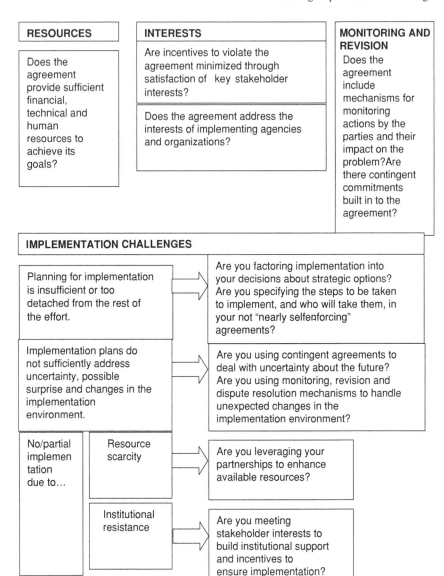

RESOURCES

Does the agreement provide sufficient financial, technical and human resources to achieve its goals?

INTERESTS

Are incentives to violate the agreement minimized through satisfaction of key stakeholder interests?

Does the agreement address the interests of implementing agencies and organizations?

MONITORING AND REVISION

Does the agreement include mechanisms for monitoring actions by the parties and their impact on the problem?Are there contingent commitments built in to the agreement?

IMPLEMENTATION CHALLENGES

Planning for implementation is insufficient or too detached from the rest of the effort.

Are you factoring implementation into your decisions about strategic options? Are you specifying the steps to be taken to implement, and who will take them, in your not "nearly selfenforcing" agreements?

Implementation plans do not sufficiently address uncertainty, possible surprise and changes in the implementation environment.

Are you using contingent agreements to deal with uncertainty about the future? Are you using monitoring, revision and dispute resolution mechanisms to handle unexpected changes in the implementation environment?

No/partial implementation due to...

Resource scarcity

Are you leveraging your partnerships to enhance available resources?

Institutional resistance

Are you meeting stakeholder interests to build institutional support and incentives to ensure implementation?

Fig. 5.1 Leveraging implementation resources

Institutional resistance may nonetheless be significant even when leaders have good incentives to support change and new resources are available. Friction can arise at the technical and field levels of an organization between "the way we do things" and "the way the crazy new program wants us to do things." There is no magic solution to this type of resistance. It requires a fairly generic set of change management skills.[12] For international negotiators, an initial step is to monitor implementing agencies on an ongoing basis, with attention both to the accomplishment of program and project goals and to the organizational dynamics that are supporting or impeding success. Devising a strategy to overcome resistance at this level can be thought of as a type of negotiation and coalition-building and analyzed and planned for in a similar manner. As with other negotiations, the first step is to identify precisely which individuals or officials within the organization will have the most influence (authoritative or informal) on the organizational outcome (in this case, implementation of an international agreement) the leader seeks to achieve. Thinking of the implementation process as a process of building a "winning coalition" can be helpful for analyzing institutional resistance and developing strategy for overcoming it.

Leaders should analyze where interests are shared or convergent, and where they might be divergent. Leaders should then develop a strategy for assuring smooth implementation through the technique of backward mapping, including identifying decision-makers, those who influence them and those who will be affected directly by decisions regarding resources. "Decision trees" can be a helpful tool to accomplish this task.

5.4 Conclusion

Implementation considerations should be part of the negotiator's focus at every stage of a negotiation, for even a widely supported agreement is worthless if it cannot be implemented. The primary tools for maximizing implementability include asking and jointly answering the key implementation questions; engaging potential implementers; anticipating and addressing contingencies; creating procedures for monitoring, revision and dispute resolution; leveraging resources through creative partnerships and clear, compelling goals; and managing organizational change and organizational resistance. (See Fig. 5.1)

Meeting implementation challenges jointly is not only a way to improve the chances that the goals of the agreement will be reached, it is also an opportunity for stakeholders to build and deepen their relationships. By meeting implementation commitments together, and addressing openly and constructively the difficulties encountered along the way, parties can continue to build trust in each other and in the value of collaborating to meet a shared public health objective.

---

[12] *See*, for example, Kotter (1996).

Chapter 6
Building Institutional Capacity for Effective Negotiation

Abstract Negotiation, by definition, requires interaction between two or more parties seeking to find a mutually acceptable agreement. The ability of negotiators to successfully achieve their interests (and those of their constituents) is dependent on their own skills, the skills of their counterparts and the support of their institution to implement the agreements reached. While negotiators may have little control over the quality of their counterparts or contextual factors, institutions do have the ability to improve—over time—their organizational effectiveness in negotiations.

Keywords Institutional capacity-building · Leadership · Assessment · Champion · Sponsor · Culture of negotiation · World Health Assembly · Gro Harlem Brundtland · FCTC · Margaret Chan · World Health Organization · UNDP Virtual Development Academy · Trade, Foreign Policy, Diplomacy and Health Unit (TFD), Department of Ethics, Trade Human Rights and Health Law (ETH), WHO · Training · Negotiation tools · Operating procedures · Incentives · Performance evaluation

Negotiation success is not only, or even primarily, a matter of choosing skilled individuals to represent the organization. It comes from putting into place a set of organizational strategies and supports for those negotiators that maximize their chances of meeting the organization's interests through the negotiation process.

Improving organizational negotiation capacity requires identifying the negotiation challenges most important to the organization's success, reflecting on how effectively the organization is meeting those challenges, revising policies and staffing that affect negotiation success and building skills to increase success. In most cases, success depends on committed leadership, a self-reflective process of assessment and a sustained program with metrics for success that can be measured by the organization's leaders.

D. Fairman et al., *Negotiating Public Health in a Globalized World*,
SpringerBriefs in Public Health, DOI: 10.1007/978-94-007-2780-9_6,
© The Author(s) 2012

6.1 Identifying Sponsors and Champions

Leadership is key for effecting institutional change. Institutional capacity for negotiation is unlikely to improve without significant leadership. Such leadership can take different forms. Sponsors at high levels of the organization (such as a Minister, Director General, etc.) are needed to approve commitment of resources—time, money, policies, rules, procedures, and people—needed to effect meaningful change (Movius and Susskind 2009: p.55). Identifying champions for a culture of negotiation, with the appropriate authority and visibility, is also important to steer change and achieve results at a more rapid pace. Champions provide operational leadership to identify problems and opportunities and design ways to integrate negotiation tools and processes into current organizational procedures and culture.

Such champions might emerge naturally. They may include, for example, former star negotiators or political leaders. Or, institutions might assign an individual to take on the role. Either way, they do not need to be top leaders; often they are in the middle of their organizations, but have the passion for negotiation and ability to lead the effort once top leaders give their blessing to the effort (Movius and Susskind 2009: p.56). Champions must recognize that institutional change is a slow process and will require resources, consensus-building and negotiation skills and persistence.

As an example, the implementation of the 1996 WHA resolution on tobacco control began only after the election of Gro Harlem Brundtland as Director-General of the WHO. Under her tenure, negotiations on the FCTC were undertaken and WHO resources made available to the Tobacco Free Initiative, which had been newly created as a Cabinet project. Brundtland was arguably the champion needed to effect institutional change within the WHO as a precondition of implementing the 1996 resolution. In a similar manner, WHO's current Director General Margaret Chan plays a leadership role in bridging differences and forging consensus on key difficult issues related to intellectual property and virus sharing and benefits sharing. Other examples of champions include the Geneva-based Ambassadors and senior government officials who chair intergovernmental meetings and working groups and who shepherd the process of consensus building within those forums. These officials are supported by senior members of the WHO secretariat, who, with the chairs and bureaus of the various intergovernmental groups develop the roadmap for successful negotiations.

6.2 Building Capacity

To create institutional capacity for effective negotiation, negotiators may need to strengthen their ability to reach and implement decisions. Skill-building opportunities—including trainings, simulations and lectures on key negotiation skills—

are widely available to institutions worldwide. From basic trainings to advanced practice sessions and coaching, capacity-building opportunities help to build confidence, skills and motivation. Very often, negotiators have a sense of what works and does not work, and ongoing training and capacity-building opportunities allow them to hone those skills and share their experiences with others. This helps to create a culture of negotiation that is shared across an institution.

For example, the United Nations Development Program offers online negotiation trainings for professional development credits as part of its Virtual Development Academy (VDA). Through the VDA, UNDP staff from around the world are able to build their skills, interact with their peers and build a common understanding of key negotiation strategies, theories and techniques. UNDP also offers in-person tailored trainings to UNDP staff and their civil society and government counterparts to help build the capacity of their own staff and their frequent negotiation counterparts.

Box 6.1 The unit on Trade, Foreign Policy, Diplomacy and Health (TFD) in the Department of Ethics, Trade, Human Rights and Health Law (ETH) works to catalyze, facilitate and coordinate actions towards greater policy coherence between the promotion and protection of health and other government policies such as trade and foreign policy. Key external partners include the World Trade Organization and UNCTAD.

  Objectives:

- To support countries in understanding and responding to the implications of international trade and trade agreements for health
- To support efforts to ensure that health is promoted and protected in the context of foreign policy
- To build the capacity of countries to negotiate in support of collective action to address global health challenges

  Areas of work:
  Foreign Policy and Global Health
  Responding to cross-border risks to public health security
  Global Health Diplomacy
  Shaping and managing the global policy environment for health
  Trade and Health
  Making trade and trade agreements work for health

Training a small group of individuals, or training lower-level staff, however, is not sufficient to build organizational capacity. Unless a critical mass of people in the organization can understand and talk meaningfully about the core concepts and processes of negotiation, institutional capacity will not likely be built. It is important that the institution develop a shared model of negotiation, and this will require that staff at all levels—including the top—understand and support for the concepts.

6.3 Developing Tools and Formal Procedures
to Institutionalize Capacity

Training and other forms of knowledge and skill-building of staff are not enough to achieve institutional change. It is impossible to train the entire staff of an organization, and a single training or course will not necessarily result in staff applying these skills in their negotiations. Participants who attend trainings need an opportunity to use the concepts *in context*. As Movius and Susskind (2009: pp.173–76) point out, people who have received training in negotiation rarely get to experiment with the skills and approaches they have learned in the high-stakes, complex, pressured situations they find themselves in at work. Moreover, colleagues often resist doing things a new way unless past experience has truly been terrible. For these reasons, further institutional support for the use of the skills and approaches of mutual gains negotiation is necessary. Institutions seeking to create a more methodical approach to negotiations can develop negotiation tools and protocols that encourage and support application of a mutual gains negotiation process throughout the organization. Simple tools such as negotiation preparation checklists, mapping worksheets, templates and suggested criteria lists can help institutions to harmonize their negotiation approaches and offer practical organizational tools for their negotiators to implement mutual gains negotiation approaches. Such tools also help institutions to maintain records of their negotiations, which may be useful for other, future negotiations. Some examples are presented in Appendix 2.

Second, institutions can offer easy access to negotiation support. This might take the form of on-call negotiation coaches (in-house or through a consultant), a library of negotiation literature, access to negotiation evaluations or ongoing training. Implementation teams can help institutions to follow through with their commitments to improve their negotiation capacity by monitoring and evaluating process, needs and innovations in the field.

Finally, the operating procedures of the institution, both formal and informal, need to be aligned to enable negotiators to be effective. Negotiation effectiveness can be undermined by barriers that are not visible, including existing procedures, mandates and incentives that create confusion or conflict about what to do, and, at worst, disincentives to use the mutual gains approaches and skills. For example, in order to develop mutual gains options, negotiators ought to have a clear mandate and autonomy to explore a number of options. They also may need sufficient time to engage in effective preparation. If the organization does not allow negotiators the time to prepare, or does not have processes in place to promote greater information-sharing and cooperation for staff of different departments (or ministries) to develop in order to work effectively as a team, negotiation effectiveness may be undermined (Movius and Susskind 2009: p.81). Leaders within organizations can create the time and space for their negotiators to adequately prepare for and evaluate negotiations. Most negotiators face competing pressures on their time and often skip important negotiation steps. Managers must be aware of these

BUILDING INSTITUTIONAL CAPACITY

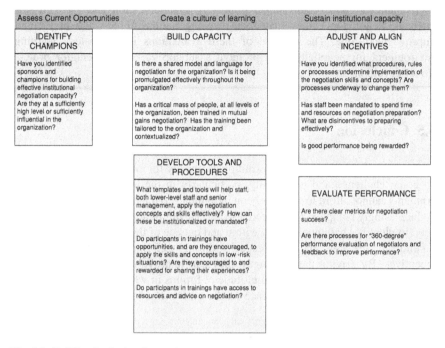

Assess Current Opportunities	Create a culture of learning	Sustain institutional capacity
IDENTIFY CHAMPIONS	**BUILD CAPACITY**	**ADJUST AND ALIGN INCENTIVES**
Have you identified sponsors and champions for building effective institutional negotiation capacity? Are they at a sufficiently high level or sufficiently influential in the organization?	Is there a shared model and language for negotiation for the organization? Is it being promulgated effectively throughout the organization? Has a critical mass of people, at all levels of the organization, been trained in mutual gains negotiation? Has the training been tailored to the organization and contextualized?	Have you identified what procedures, rules or processes undermine implementation of the negotiation skills and concepts? Are processes underway to change them? Has staff been mandated to spend time and resources on negotiation preparation? What are disincentives to preparing effectively? Is good performance being rewarded?
	DEVELOP TOOLS AND PROCEDURES	**EVALUATE PERFORMANCE**
	What templates and tools will help staff, both lower-level staff and senior management, apply the negotiation concepts and skills effectively? How can these be institutionalized or mandated? Do participants in trainings have opportunities, and are they encouraged, to apply the skills and concepts in low -risk situations? Are they encouraged to and rewarded for sharing their experiences? Do participants in trainings have access to resources and advice on negotiation?	Are there clear metrics for negotiation success? Are there processes for "360-degree" performance evaluation of negotiators and feedback to improve performance?

Fig. 6.1 Building institutional capacity

dynamics and create opportunities for their negotiators to spend the time needed to prepare for negotiations using mutual gains concepts and applying negotiation best practices.

6.4 Evaluating Performance

Once negotiation is identified as a key competency area and staff have been given adequate opportunities to build their skills, institutional incentives for use and ongoing improvement of these competencies are necessary to ensure that the organization's capacity in negotiation is built. Institutions can, for example, incorporate negotiation skills and achievements into their employee professional development or evaluation processes, in order to signal the importance of nego-tiation as a professional competency and encourage staff to develop these skills in order to advance in the organization. After a negotiation, for example, an insti-tution might request a "360 degree assessment" (i.e., feedback from superiors, subordinates and peers) of the negotiator's performance and use that assessment to offer constructive feedback for improvement.

An institution should also develop its own metrics for negotiation success, such as maintained good relationships, improved communication and satisfied key interests. Negotiators can evaluate their own negotiations to see where they might improve next time. The purpose of these evaluations is to create a learning environment for negotiators and promote a supportive environment for professional development.

6.5 Conclusion

As Fig. 6.1 illustrates, building institutional capacity for effective negotiation is an on-going challenge and a necessity, particularly in the developing world. Governments cannot afford to rely solely on individuals to manage their negotiations in an increasingly complex, multifaceted world. Health policy-makers and advocates need to identify champions—both within and outside of their organizations, create a culture of learning, and build capacity through offering appropriate tools and incentives. By driving these changes, they can improve negotiation preparation, performance and learning in the multitude of forums in which they will negotiate, and ultimately gain greater influence on the outcomes of their negotiations.

Part II
Case Studies

Chapter 7
Case I—Analyzing a Complex Multilateral Negotiation: The TRIPS Public Health Declaration

Abstract On November 14, 2001, trade ministers from around the world approved the Declaration on the TRIPS Agreement and Public Health ("Public Health Declaration") at the end of the Fourth World Trade Organization (WTO) Ministerial Conference in Doha, Qatar. The Public Health Declaration was the product of months of negotiations—described in this case—examining the Agreement on Trade-Related Aspects of Intellectual Property Rights ("TRIPS Agreement") and its impact on the public health sector. Developing nations, confronted with emergencies like the AIDS crisis, encouraged the adoption of an approach to the TRIPS Agreement that would provide them with a greater degree of flexibility when dealing with matters related to public health. A bloc of developed nations, led by the United States, urged a more cautious reading of the TRIPS Agreement, emphasizing that patent protection was necessary for providing incentives to create new drugs. The Public Health Declaration reflected a consensus view that the TRIPS Agreement should be construed by governments in a way that supports the realization of their public health goals.

Keywords TRIPS · WTO · Doha Public Health Declaration · Negotiations · Issue framing · Coalition building · NGOs · Access to essential medicines · HIV/AIDS · Preparation for negotiation · Africa Group · Interests

---

Case study researched and written by C. Michel Roh.

D. Fairman et al., *Negotiating Public Health in a Globalized World*,
SpringerBriefs in Public Health, DOI: 10.1007/978-94-007-2780-9_7,
© The Author(s) 2012

7.1 Introduction to the Agreement

The Public Health Declaration itself is composed of seven paragraphs.[1] Among its more important provisions, it confirms the right of a government to resort to parallel importation and compulsory licensing.[2] *Parallel importation* occurs when the buyer of a patented product exports the product for resale to a country where the patent holder charges a higher price.[3] A government that practices *compulsory licensing* may permit a manufacturer to produce and sell a patented product without the consent of the patent holder.[4] Both compulsory licensing and parallel importation are tools that governments can use to provide cheaper access to medicines.

Arguably the most important provision of the Public Health Declaration, from the perspective of developing countries, is in Paragraph 4. This paragraph states that the "TRIPS Agreement does not and should not prevent members from taking measures to protect public health."[5] Although the paragraph avoids the more binding phrase "shall not," the text still gives broad support to the notion that governments can do whatever they find necessary to protect the health of their citizens. Intellectual property rights are thus not presented as insurmountable obstacles to the treatment of sick individuals in impoverished regions of the world.

On the other hand, the Public Health Declaration does not repudiate all intellectual property rights. Paragraph 3 presents recognition that "intellectual property protection is important for the development of new medicines."[6] Developed countries therefore had one of their interests met: The Public Health Declaration could not be seen as a direct attack on the integrity of the TRIPS Agreement.

The Public Health Declaration does leave some questions unanswered. For instance, Paragraph 6 leaves unresolved the question of how countries lacking the manufacturing capacity to produce drugs on their own can take advantage of the compulsory licensing provisions.[7] Furthermore, there are many problems obtaining access to medicines that are unrelated to intellectual property rights or the TRIPS Agreement. Medical experts have recognized, for example, that without improved infrastructure and better means of distribution, medicines will not get to

---

[1] A complete text of the Public Health Declaration can be found at http://www.wto.org/english/thewto_e/minist_e/min01_e/mindecl_trips_e.htm.

[2] See *id.*, paragraphs 5b and 5d.

[3] See Alan O. Sykes, "Public Health and International Law: TRIPS, Pharmaceuticals, Developing Countries, and the Doha 'Solution,'" *Chicago Journal of International Law*, Volume 3, Spring 2002, p. 63.

[4] For a broader definition of compulsory licensing, see Ellen 't Hoen, "Public Health and International Law: TRIPS, Pharmaceutical Patents, and Access to Essential Medicines: A Long Way from Seattle to Doha," *Chicago Journal of International Law*, Volume 3, p. 32.

[5] See Public Health Declaration, *supra* note 1, at paragraph 4.

[6] *Id.*, paragraph 3.

[7] *Id.*, paragraph 6.

those who need them no matter what the price.[8] The Public Health Declaration could not even attempt to resolve these types of issues, since they remain beyond the scope of the TRIPS Agreement.

Many people have nevertheless represented the Public Health Declaration as a victory for developing countries in their efforts to gain access to essential medicines.[9] The Public Health Declaration confirmed that these countries have a variety of alternative means to obtain needed medicines. Such countries are thus in an improved position in their negotiations with the pharmaceutical industry to obtain essential medicines and have legitimate alternatives if such negotiations do not come to fruition.

7.2 Introduction to the Public Health Declaration Negotiation Process

The Public Health Declaration was not just handed to developing countries. Developed countries like the United States were wary of interpreting the TRIPS Agreement in a way that would reduce the rights of patent holders.[10] Such countries characteristically do not have the health problems faced by many poor nations. In addition, they have the interests of their pharmaceutical industry to

---

[8] Médecins sans Frontières lists the following factors as having influence on "access to effective medicines: quality of diagnosis; accurate prescribing, selection, distribution and dispensing of medicines; drug quality; capacities of health systems and budgets; lack of research and development (R&D); and price." See http://www.accessmed-msf.org/campaign/faq.shtm.

[9] See, e.g., Ellen 't Hoen of Médecins sans Frontières who stated that "[the Public Health D]eclaration is a major step forwards in the quest to ensure access to medicines for all." James Love of Consumer Project on Technology said that "[the Public Health Declaration] is the strongest and most important international statement yet on the need to refashion national patent laws to protect public health interests." But see the comment of Asia Russell of Health Gap Coalition and Act Up Philly. She said that "[d]eveloping countries came to Doha to extract a clear declaration that public health and access to medicines are more important than protecting the commercial interests of pharmaceutical companies. At the end of the day, opposition from rich countries crippled the legally binding language sought by the majority of WTO countries." For all of these quotations, see "Views on the Draft Declaration on the TRIPS Agreement and Public Health," at http://www.cptech.org/ip/wto/doha/ngos11132001.html, which provides statements that various NGO leaders made immediately after the Public Health Declaration was released.

[10] For instance, even relatively late in the negotiation process, the American trade delegation was emphasizing that "[e]ffective patent systems, therefore, were crucial to finding better treatments and, ultimately, cures for HIV/AIDS and the many other diseases and health conditions that afflicted the world's population. The TRIPS Agreement appropriately required Members to provide such effective patent systems." Minutes of the September 19–20, 2001, Meeting of the Council for Trade Related Aspects of Intellectual Property Rights, ¶165, ref. no. IP/C/M/33 available through http://docsonline.wto.org/gen_trade.asp.

keep in mind.[11] A declaration acknowledging exceptions to patent protection clearly goes against the financial interests of these companies. Due to these competing interests, a final agreement could only be achieved through a complex negotiation process.

7.2.1 Public Health Outside the Framework of the Doha Negotiations

The beginnings of this process did not involve country-to-country negotiations. Civil society first had to engage in a long campaign to raise awareness. During the 1990s, nongovernmental organizations (NGOs) like Médecins sans Frontières (MSF), Health Action International and Consumer Project on Technology engaged in a lobbying effort aimed at ensuring cheaper prices to essential medicines for developing nations in need. Notably, these organizations arranged a conference in Geneva in March 1999 at which representatives of the pharmaceutical industry, various NGOs, national governments and intergovernmental organizations came together to discuss the potential for compulsory licensing under Article 31 of the TRIPS Agreement.[12] At the end of that year, the MSF officially began its Access to Medicines Campaign,[13] a campaign that had added public significance since MSF won the 1999 Nobel Peace Prize.

By the middle of 2001, official policies were beginning to show the effects of the access to medicines campaign.[14] In 1999, Charlene Barshefsky, then the United States Trade Representative, agreed to back South Africa's effort to obtain medicines at more affordable prices as long as South Africa did not violate American patent law.[15] In April 2001, 39 drug companies agreed to drop their lawsuit protesting legislation in South Africa that limited their patent rights.[16]

---

[11] For a sampling of the perspective of the pharmaceutical industry, see PhRMA, "Frequently Asked Questions" in "Health Care in the Developing World" at http://world.phrma.org/faq.html.

[12] For information concerning this meeting see Consumer Project on Technology, "March 1999 Meeting on Compulsory Licensing of Essential Medical Technologies" at http://www.cptech.org/march99-cl/.

[13] For information regarding MSF's Access to Medicines campaign see MSF, "Campaign for Access to Essential Medicines" at http://www.accessmed-msf.org/index.asp.

[14] Unless otherwise specified, I use "access to medicines campaign" as a general name for the popular movement led by various NGOs to provide affordable medicines to all people, particularly those in developing nations.

[15] Ceci Connolly, "U.S., South Africa Reach Deal on AIDS Drug Sales," *Washington Post*, September 18, 1999, at A11, available at http://jobs.washingtonpost.com/wp-srv/politics/campaigns/wh2000/stories/aids091899.htm.

[16] See "Drug Companies Drop Lawsuit Against South Africa," *USA Today*, April 19, 2001 at http://www.usatoday.com/news/world/2001-04-19-drugsuit.htm.

Moreover, drug companies had entered into several agreements to provide quantities of necessary medicines to developing countries at affordable prices.[17]

The extent to which the access to medicines campaign began to affect industry and government decision-making can best be measured by the United States' decision to refrain from pursuing a WTO action against Brazil based upon the latter's patent law. The United States believed that Article 68 of Brazil's law had nothing to do with access to medicines but was merely a measure destined to "create jobs for Brazilian nationals."[18] Brazil, on the other hand, considered the patent law vital to its efforts to make AIDS drugs available to Brazilians.[19] Based on each side's original positions, it is not easy to see how this matter could be resolved before the end of the adjudication process.[20] However, the U.S. government dropped the WTO case only months after issuing the above statement (removing it to a bilateral and more informal dispute resolution process). Robert Zoellick, who served as U.S. Trade Representative from 2001–2005, stressed that the United States was willing to work with Brazil "toward our shared goal of combating the spread of this dangerous virus [AIDS]."[21] One can suppose that the negative publicity the United States faced and the strong policy justifications that Brazil could advance made the United States reevaluate its pursuit of this WTO action.

7.2.2 WTO Public Health Negotiations in the Months Before Doha

Within this context, formal negotiations on the public health issue began in the WTO TRIPS Council in early 2001. In March 2001, the Africa Group requested a formal session of the Council to explore the relationship of the TRIPS Agreement to public health. Over the coming six months, the issue would be discussed at two

---

[17] See, e.g., Rachel Zimmerman, "GlaxoSmithKline Plans to Announce Cut In Prices for AIDS Drugs to Poor Countries," *Wall Street Journal Europe*, June 11, 2001, at http://www.accessmed-msf.org/prod/publications.asp?scntid=318200182119&contenttype=PARA&; Paul Blustein and Barton Gellman, "HIV Drug Prices Cut for Poorer Countries: Other Firms May Follow Merck's Lead," *Washington Post*, March 8, 2001, A1, available at http://www.washingtonpost.com/ac2/wp-dyn?pagename=article&node=&contentId=A38407-2001Mar7&notFound=true.

[18] USTR, "2001 Special 301 Report," available at http://www.ustr.gov/enforcement/special.pdf.

[19] "Statement of Jose Serra, Minister of Health, to the 2001 USTR Special 301 Report," May 3, 2001, available at http://www.cptech.org/ip/health/c/brazil/serra05032001.html ("The production of drugs to control AIDS helps us to save as much as $US200 million per year on purchases from abroad … This [AIDS] programme is what it is thanks to the determination of the government of Fernando Henrique Cardoso to bring down the costs of these drugs.")

[20] For instance, see *Id.* ("There is no way that the Brazilian Government will retreat on this issue.")

[21] "United States and Brazil agree to use newly created Consultative Mechanism to promote cooperation on HIV/AIDS and address WTO patent dispute," Office of the United States Trade Representative, June 25, 2001, available at http://www.ustr.gov/releases/2001/06/01-46.htm.

formal meetings (in June and September 2001) and one informal meeting (in July 2001). At these meetings, countries set forth their positions and introduced official documents. The Africa Group, the European Union and a number of countries aligned with the United States all took advantage of the TRIPS Council setting to introduce papers in support of their reading of the TRIPS Agreement.[22]

As 2001 continued, support for a separate TRIPS/Public Health Declaration grew. By the time the Doha Ministerial Meeting was being prepared, it was clear that a draft declaration was necessary. The chairman of the General Council of the WTO produced such a draft, which then formed the basis of discussion at Doha.[23] The major points of contention concerned the circumstances under which a country could take advantage of the flexibilities offered by the TRIPS Agreement, and the fundamental relationship between the TRIPS Agreement and public health. Did public health concerns always preempt intellectual property rights? This question had to be answered at the Doha meeting itself.

Trade delegations managed to surmount these problems while at Doha. Small groups of country representatives gathered to hammer out the details of the compromises that needed to be reached. However, countries only agreed to the final text at the proverbial eleventh-hour. On November 14, 2001, just prior to the close of the Ministerial Conference, WTO trade ministers finally ratified the Public Health Declaration.

The above paragraphs set out a timeline of the major events in the TRIPS/ Public Health negotiations. The rest of this paper will present an analysis of the negotiating process itself and the effects of that process on the outcome. It will mainly concentrate on the months leading up to Doha, and on the governments that participated in the negotiations. The focus will be on how developing countries managed to obtain a declaration that reflected many of their interests and to see if lessons can be taken that can be applied in other fora.

---

[22] It is unlikely that much of the negotiation process actually took place during the Council meetings. Because a public record was maintained, there would be too many opportunities to lose face through inopportune comments or inconvenient concessions. Therefore, any statement made at a meeting had to be guarded—a delegation's initial statement was most likely scripted to the largest extent possible. Furthermore, the issues were of enough importance that no participant at these meetings could realistically bind their delegation to any changes proposed on the spot (even presuming that they had such authority, which was probably not the case in most instances). Each delegation would want time to consider every proposal made. Therefore, throughout the Doha negotiations, there must have been numerous informal discussions among the various negotiating parties. It could only have been in this manner that coalitions were solidified and progress was made in the negotiations.

[23] General Council, "Draft Declaration on Intellectual Property and [Access to Medicines] [Public Health]," October 27, 2001, available at http://www.ictsd.org/ministerial/doha/docs/ IP27oct.pdf. It is difficult to know if another draft was circulated between October 27, 2001, and the beginning of the Doha conference on November 9. My research has not discovered another draft between these dates. Regardless, the conclusions of this paper would not be changed if there was another draft.

7.3 Participants/Coalitions

The Doha negotiations involved hundreds of participants. The numerous countries that are party to the TRIPS Agreement would be bound by any eventual declaration, to the extent that such a declaration was legally enforceable. Each member country thus had an official trade minister who served as the chief representative of their government at Doha.

Of course, in many instances these officials could not actively involve themselves in the minutia of the public health negotiations, since being a trade minister involves many obligations. At Doha, public health was one of many issues being discussed. Therefore, particularly among the richer nations, it was common practice to send a large delegation composed of multiple trade officials.[24] The lower-level officials could specialize in certain aspects of the issues under consideration at Doha, freeing the trade minister to deal with broader concerns.

In addition to the official government delegations, many members of civil society assisted in the negotiating process. These participants would not, of course, have a final say on whether a particular agreement was accepted or not. However, numerous NGOs did attend the Doha negotiations.[25] These NGOs used written communications to publicize and comment on the negotiating positions taken by various countries, in the hope of influencing the outcome of the negotiations.[26]

The nature of the public health negotiations necessitated a redistribution of roles. Not everyone could have their voice heard in the negotiating room. Impoverished developing countries, like many African nations, could not afford to

---

[24] For instance, at Doha the EU delegation was composed of 502 representatives, the United States sent 50 officials, Japan had 168 representatives and the Canadian delegation numbered 62. Meanwhile poorer nations like Mali, Sri Lanka, Sierra Leone and Jamaica could only send between 2 and 4 representatives. Sabrina Varna, "Doha: A Case of Bad Process and Good PR?" in *South Bulletin* (a South Centre publication), Bulletin 24–25, November 30, 2001, available at http://www.southcentre.org/info/southbulletin/bulletin24-25/bulletin24-25-07.htm. For a complete list of participants, *see* "List of Representatives," WTO, WT/MIN(01)/INF/15/Rev.1, December 11, 2001, at http://docsonline.wto.org/DDFDocuments/t/WT/Min01/INF15R1.doc. The numbers given by Ms. Varna do not exactly correspond to those listed by the WTO, at least with respect to the United States. Nevertheless, Ms. Varna was correct in noting the great discrepancy in the number of trade delegates various countries sent to Doha.

[25] *See* http://www.wto.org/english/thewto_e/minist_e/min01_e/doha_attend_e.doc for a complete list of the NGOs that were present at the Doha negotiations. Only one representative of each NGO was permitted to attend. *See* "647 non-governmental organizations eligible to attend the Doha ministerial," *WTO News: 2001 Press Releases*, August 13, 2001, available at http://www.wto.org/english/news_e/pres01_e/pr240_e.htm

[26] *See, e.g.*, James Love, "Letter to USTR Zoellick regarding WTO Patent Discussions," November 10, 2001, available at http://www.cptech.org/ip/wto/doha/lovezoel11102001.html .James Love is Director at the Consumer Project on Technology. *See also* "TRIPS: Will the majority prevail," November 11, 2001, available at http://www.cptech.org/ip/wto/doha/ngos11112001.html. This letter was issued jointly by Act-Up Paris, Consumer Project on Technology, Consumers International, Health GAP Coalition, MSF, Oxfam, Tebtebba Foundation and Third World Network.

have large trade delegations and individual representatives at every public health meeting.[27] And, it would be impossible to take into account the hundreds of separate and often contrasting opinions in the writing of one short declaration. The different nuances in language that each delegation might want would require endless negotiating.

This problem had two possible resolutions. Either some parties could withdraw from the negotiation process and just agree to adhere to the result, or these same parties could band together into factions and choose representatives to support their positions. Most developing countries chose to overcome their problem of negotiating capacity by banding together.[28] The developing country coalition sought a more liberal reading of the TRIPS Agreement in the effort to increase access to medicines; it was headed by such countries as Brazil, India and Zimbabwe (of the larger group of African nations known as the Africa Group). At the same time, pharmaceutical-producing nations such as the United States and Switzerland supported each other in an effort to ensure that pharmaceutical patent rights were not ignored.

To these two (unofficial) coalitions must be added two other factions. First, the NGOs clearly made their presence felt throughout the negotiating process, including while at Doha.[29] With their media savvy,[30] they tried to steer the negotiations in the direction they thought best. Second, certain parties did not play an active role in lobbying for either of the two proposed two solutions.[31] Most notably, the EU did not take a strong position in the debate, either in the lead-up to Doha or at the Doha meetings themselves. Although the EU did submit a paper to the initial TRIPS Council Meeting dedicated to public health,[32] they did not endorse either of the competing Public Health Declaration positions in the lead-up

---

[27] See supra note 24. Botwana sent 16 delegates; Burundi had 3 representatives; Cameroon had a delegation of 10. See "List of Representatives," supra note 24. However, it is not fair to claim that all African nations sent small delegations. Nigeria's delegation numbered around 40, which is comparable to the size of the American delegation.

[28] It is important to differentiate between the smaller official coalitions, such as the Africa Group and ASEAN, and the larger unofficial coalition that was formed to negotiate for improved access to medicines. Brazil, while one of the leaders of the unofficial coalition, was not part of either the Africa Group or the Association of Southeast Asian Nations (ASEAN). Furthermore, no NGO had membership in any of the official coalitions.

[29] For instance, at Doha some NGOs engaged in anti-globalization protests. See C. Rammanohar Reddy, "Why the poor love the E.U., U.S." The Hindu (online edition), November 12, 2001, available at http://www.hinduonnet.com/thehindu/2001/11/12/stories/0512134e.htm.

[30] The NGOs have been particularly strong at entering the debate through the medium of the Internet. For examples, see www.cptech.org, www.twnside.org, and http://www.accessmed-msf.org/index.asp. Each of these Internet sites provides detailed advocacy of their positions in the access to medicines campaign.

[31] For instance, South American countries, apart from Brazil, did not play a major role in these negotiations.

[32] See "The relationship between the provisions of the TRIPS Agreement and access to medicines," Communication from the European Communities and their member states, June 11, 2001, available at http://www.wto.org/english/tratop_e/trips_e/paper_eu_w280_e.htm.

to Doha. There was simply a difference in opinion among their member states about what position the EU should take.[33] The EU instead sought out a compromise solution based on a tiered-pricing scheme.[34] Perhaps the role of the EU in the Doha negotiations can thus be best categorized as that of a broker between the opposing sides, although at least one author even discounts this suggestion.[35]

The dangers of negotiating through coalitions are twofold. First, the leaders of a coalition might not adequately address the interests of certain members of the coalition. Second, certain members might act unilaterally against the will of the larger coalition, thereby fracturing the group. In the Doha case, the developing nation coalition was in greater danger of succumbing to these ills.

7.3.1 Pressures on the Developing Nation Coalition

The developing nation coalition was composed of more than 60 nations from three continents. Some of the countries involved, including India and Brazil, were producers of generic drugs. These countries perhaps had different interests than those facing the AIDS crisis without any significant pharmaceutical manufacturing capacity and with a lesser degree of access to AIDS medicines. Developing countries also differed depending upon the degree to which AIDS and other pandemics were actually affecting their populations. In addition, certain non-producers of generic pharmaceuticals may have seen the negotiations as a first step in the development of such an industry. It was therefore possible that there might be ways of peeling off certain countries from the coalition by addressing the interests of individual countries or perhaps the interests of one particular region of the world.

Furthermore, the developing nation coalition was being supported by NGOs who did not necessarily share the countries' agenda. NGOs do not have to deal

---

[33] *See* Sukumar Muralidharan, "A compromise deal," *Frontline*, Volume 18, Issue 24, Nov. 24–Dec. 7, 2001, available at http://www.flonnet.com/fl1824/18240140.htm. ("The European Union (E.U.) was anxious to show sensitivity, although it remained torn by conflicting perceptions among its member-states").

[34] *Id.* ("The halfway solution [the E.U.] proposed was to establish a system of tiered (or differential) pricing, which would encourage drug manufacturers to sell at relatively lower prices to the poorer countries. But this, E.U. spokespersons said, would require a reciprocal commitment from developing countries to bar the re-export of drugs obtained at concessional prices.")

[35] *See* Martin H. Godel, *The Doha Conference: Birth of a New Trade Round*, (Masters Thesis prepared for The Fletcher School of Law and Diplomacy), April 26, 2002, page 52, available at http://nils.lib.tufts.edu/Fletcher/MartinGodel.pdf ("The fact that the EU did not play a very active role regarding TRIPS – neither as a participant nor as a broker ..."). Nevertheless, it is important to note that the EU was a member of the Working Group at Doha that hammered out the final agreement. *See* Alan Larson, "A New Negotiating Dynamic at Doha" (available through the U.S. embassy in China), available at http://www.usembassy-china.org.cn/press/release/2002/0302e-Alan%20Larson-Doha.html.

directly with the daily problems of governing, and they can focus on a certain problem, like access to medicines, to the exclusion of all other issues.[36] Governments, by contrast, must confront many distinct problems, knowing that a decision to concentrate on one issue may mean that another issue does not receive the resources it requires. Developing countries also rely on developed countries for aid, and therefore often cannot afford to disrupt relations with donor countries. Developing countries thus had to decide what priority to give the access to medicines campaign, whereas NGOs dedicated to the issue did not have to make that decision.

There was therefore the distinct possibility that the developing nation coalition could fracture due to exterior pressures from NGOs. If certain developing countries decided that they could afford to dedicate resources to the access to medicines problem, they might be willing to demand the concessions that the NGOs felt were required for an optimal resolution of the issue. Other developing nations might have felt as if they could not afford to take a stand on this issue, because they required concessions or aid in another sector.

Despite these pressures, the developing nation coalition maintained its unity throughout the negotiations. Certainly, countries like Angola must have realized that to stray outside of the coalition would make them a non-factor in the negotiations. The United States would not pay attention to countries like Angola if they acted independently, as these countries have no leverage.

It is initially less clear how Brazil and India managed to cooperate with the Africa Group. After all, the member states of the Africa Group were not in a position to produce pharmaceuticals. However, Brazil and India are known for producing pharmaceuticals at a cheaper price than that otherwise charged by American and Swiss pharmaceutical manufacturers.[37] It was therefore in the interest of the African nations to see Brazilian and Indian companies produce and export many of the AIDS and tuberculosis drugs.

Another reason the developing nation coalition did not disintegrate is the strong leadership within the trade delegations of Brazil, India and the Africa Group. By the time the Doha Ministerial meetings took place, Brazil had already shown its power by standing up to the United States in its trade dispute over patent rights.[38] India's leadership was also widely recognized.[39] And, Ambassador Boniface

---

[36] An NGO can of course focus on more than one issue.

[37] *See* Johanna McGeary, "Paying for AIDS Cocktails: Who Should Pick Up the Tab for the Third World," *Time* (2001), available at http://www.time.com/time/2001/aidsinafrica/drugs.html ("India and Brazil have vigorously exploited a time lag until international patent rules apply to them, manufacturing copies of AIDS drugs and selling them at deeply discounted prices.")

[38] *See supra* pages 5–6.

[39] *See, e.g.*, William Drozdiak and Paul Blustein, "Serious Conflicts Threaten Trade Talks: Battle of Rich, Poor Nations Could Kill Planning for New WTO Discussion," *Washington Post*, November 6, 2001, at A1, available at http://www.washingtonpost.com/ac2/wp-dyn?pagename= article&node=&contentId=A44856-2001Nov5&notFound=true; Sukumar Muralidharan, "A compromise deal," *supra* note 33.

Chidyausiku of Zimbabwe was the Chair of the TRIPS Council throughout its initial meetings in 2001 on TRIPS and public health. The developing nation coalition was thus blessed with leaders who understood their position well and would not buckle under diplomatic pressure. Such strong leadership must have given confidence to other members of the coalition and made them less willing to abandon the effort.

Finally, the NGOs posed less of a threat to fracture the developing country coalition due to the fact that the coalition was the only vehicle through which they could see their goals achieved. NGOs were not of course official parties to the negotiation. They therefore had to rely on developing countries to push forward the agenda they supported. And, the success of developing countries depended in large part on maintaining the unity of the coalition. As noted above, a single developing nation, like an NGO, had little power to affect the ultimate outcome of the negotiations.[40] It was only when they were united that they could push forward their agenda. Thus, we find that when the negotiations were hitting their most tense moments, NGOs were proclaiming their strong support for the proposals of developing countries.[41]

In the end, then, the developing nation coalition maintained its sense of direction. The common interests, strong leadership and assistance from civil society made this coalition a strong negotiating bloc.

7.3.2   The U.S.-Led Coalition

By contrast, the coalition supporting a more cautious reading of the TRIPS Agreement was in greater disarray. This coalition was composed of only a small number of countries, the most important of which were the United States, Canada, Switzerland, Australia and Japan. While the United States obviously could wield an enormous amount of leverage due to its economic, political and military power, its coalition was in the delicate spot of opposing a position that was publicly supported by a far larger number of countries. The U.S.-led coalition was thus vulnerable to attacks claiming that it was obstructing the will of the world.[42]

The other problem for the coalition was that its major supporter in civil society was the pharmaceutical industry. This industry is very unpopular in certain circles, owing to the fact that pharmaceutical companies make large profits on drugs that

---

[40] *See supra* page 13.

[41] "TRIPS: Will the majority prevail?" *supra* note 26. ("We urge the developed countries, particularly the United States, Japan, and Switzerland, to withdraw their opposition to developing country proposals.")

[42] *See id.*

save lives.[43] To many people, it does not seem fair to allow pharmaceutical companies to capitalize on their patents, without forcing them to make these drugs available to save the lives of those who cannot otherwise afford them.[44] The United States may not have agreed with the position adopted by the NGOs leading the charge in the access to medicines debate. However, the U.S. also likely felt that the support of the pharmaceutical industry was not of great assistance in capturing the hearts and minds of the general population.[45] Indeed, the stark contrast between the economic success of the pharmaceutical companies[46] and the incredible number of AIDS-related deaths must have damaged the cause of the American-led coalition.[47]

Thus, the American-led coalition had distinctly less moral authority than that of the developing nation coalition. Nevertheless, the sheer power of the United States, as well as the wealth and resources of its negotiating partners, left these countries in a position where under ordinary circumstances they might have been able to achieve all of the concessions they desired.[48]

---

[43] For an example of this attitude, see "WTO Declaration on TRIPS and Health: People With AIDS 1, Drug Industry 0," ACT-UP Paris, November 14, 2001, available at http:// www.cptech.org/ip/wto/doha/actupparis11152001.html. In this article, the author notes that "the dogma of corporate monopoly on life-saving drugs is no longer law."

[44] The argument is perhaps most explicitly made in the March 28, 2001, letter of Ralph Nader, James Love and Robert Weissman to Secretary Tommy Thompson of the U.S. Department of Health and Human Services, available at http://www.cptech.org/ip/health/econ/ CPTthompson03282001.html, in which they write that "[I]t is morally repugnant for the U.S. government to permit private parties to obtain exclusive rights to market these [drug] inventions in South Africa, the Philippines, Brazil, Kenya, Romania and other countries, without provisions to help make these products available to save the lives of poor and middle class people."

[45] To be fair, there is a moral argument in favor of the maintenance of patents. It is argued that patents, and the profits that are derived from them, do provide the incentive to increase medical research, which can then lead to more breakthroughs. For instance, see "USTR Zoellick Says World Has Chosen Path of Hope, Openness, Development and Growth," November 14, 2001 USTR press release, available at http://www.ustr.gov/releases/2001/11/01-100.htm ("And [the Public Health Declaration] also recognizes the importance of intellectual property protection for the development of new lifesaving medicines.")

[46] For instance, GlaxoSmithKline reported total after-tax profits of 3,190 million pounds. See "Announcement of Annual Results 2001," GlaxoSmithKline, February 14, 2002, available through http://www.gsk.com/financial/resultsannual2001.htm. GlaxoSmithKline produces HIV therapies. Id.

[47] It is estimated that in Mozambique alone, 114,111 people died of AIDS in 2001. See "HIV/ AIDS–deaths" in The World Factbook, Washington, D.C.: Central Intelligence Agency, 2002; Bartleby.com, 2002, available at http://www.bartleby.com/151/a33.html. Of course, it would be wrong to claim that if GlaxoSmithKline had eliminated its profits in the effort to produce AIDS medications, all of these lives would have been saved.

[48] See discussion infra page 30.

7.4 External Influences

Outside events contributed to make sure that countries like the United States would not be able to harness all of their material advantages. Indeed, the Doha negotiations took place in an atmosphere greatly favorable to the position of the developing countries. Developing countries saw their leverage increase due to (i) the urgent need to combat the AIDS epidemic, (ii) the September 11 terrorist attacks, (iii) the anthrax attacks, and (iv) the need for a successful end to the Doha Ministerial.

The AIDS problem was, and remains, a crisis of gigantic proportions. As many as 18 million people are expected to die of AIDS by 2010.[49] The gravity of the situation is unlike any other the world is facing today. The issue is literally one of life and death. Many people will die unnecessarily unless something is done in the very short term.

The gravity and nature of the crisis had to affect the negotiators. Most notably, they had to appear as if they were doing their part to combat the crisis. The developing countries did not have such a difficult time making their case. After all, they were seeking broad measures that would allow them to acquire necessary medicines more easily and cheaply. The developed countries' defense of patent rights was more difficult to harmonize with an attempt to appear responsive to the AIDS crisis. Patent rights enable drug manufacturers to establish a monopoly on life-saving drugs and sell them at whatever price they choose. The American-led coalition had to emphasize its interest in responding to the AIDS crisis, along with other pandemics,[50] and they had to develop arguments as to why patent rights were not part of the problem, but rather part of the solution.[51]

This approach was difficult to navigate. Such arguments would lead to the rejection of a potential method of attacking the AIDS crisis (i.e., the softening of patent rights) without giving this method an opportunity to succeed. Under certain circumstances, that would not be a great cause of worry. Not every potential solution should be tested in practice. However, the solution of reducing patent rights was controversial precisely because it was one in which many governments, civilians and NGOs believed. Due to the gravity of the AIDS crisis, all but the most diehard supporters of patent rights must have found it difficult to negotiate to

---

[49] *See* UNAIDS Fact Sheet, "AIDS and population" June 2000, available at http://www. unaids.org/fact_sheets/files/Demographic_Eng.html.

[50] *See* "Preambular language for ministerial declaration" (Contribution from Australia, Canada, Japan Switzerland and the United States), IP/C/W/313, October 4, 2001, available at http:// www.wto.org/english/tratop_e/trips_e/mindecdraft_w313_e.htm.

[51] *See* "Reply from DHHS Secretary Tommy Thompson to Ralph Nader," July 6, 2001, available at http://www.cptech.org/ip/health/econ/thomnade07062001.html ("The massive challenges nations in sub-Saharan Africa and elsewhere face in containing the spread of HIV/AIDS, tuberculosis, and malaria are not created by patent systems or by patents, and solutions do not lie in the area of patent policy.") ("[Imposing certain licensing conditions on patents] would lead to fewer innovative medicines and therapies reaching patients around the world.")

foreclose the possibility of using this potential weapon against the AIDS crisis. After all, a mistake on their part could cost millions of lives in the short term.

The September 11, 2001 terrorist attacks and their aftermath also affected the negotiations. For starters, countries like the United States reduced the size of the trade delegations they sent to Doha.[52] And, the September 11 attacks created a new world atmosphere. The United States, in particular, was seen as more vulnerable than in the past. It is impossible to measure the impact of this new vulnerability on the negotiations, or even to know whether it made the negotiations easier or more difficult than otherwise. It is just important to note that the September 11 attacks, as well as the security threat in Doha, loomed over the negotiations.[53]

The anthrax crisis of October 2001 had a more measurable impact on the negotiations. When the Americans and Canadians began to fear bioterrorism, they discovered that their stocks of antibiotics were not sufficient to combat a sustained epidemic of anthrax. In each country a debate emerged over whether a compulsory license for Ciprofloxacin should be issued, to the detriment of the patent holder Bayer. Canada eventually issued such a license, while the U.S. chose not to.[54] This controversy emerged after the United States and others had long sought to limit the circumstances and conditions under which compulsory licenses could be issued. The situation thus aided the negotiators from developing countries. It had to be difficult for American and Canadian negotiators to caution governments that compulsory licensing was only to be used as a last resort, when these countries themselves had issued or discussed issuing compulsory licenses immediately upon feeling threatened by a disease they felt ill-prepared to combat.[55]

Lastly, one non-health-related issue also put pressure on the American negotiators. The failure of the Seattle WTO Ministerial Conference is well known. The WTO and First World free-trade advocates like the United States could not afford to have another failure in Doha. It would have hurt the prestige of the WTO too greatly

[52] "U.S. plans scaled back delegation after meeting fixed for Doha," *Inside U.S. Trade*, October 26, 2001, available at http://www.globalexchange.org/wto/insideUSTrade102601.html.

[53] *See id.*

[54] *See generally* Amy Harmon & Robert Pear, "Canada Overrides Patent for Cipro to Treat Anthrax," *New York Times*, October 19, 2001, available at http://www.globalaging.org/health/world/canadaanthraxpatent.htm. This article discusses both Canada's decision and American threats to override the Bayer patent.

[55] *See* Sabin Russell, "U.S. push for cheap Cipro haunts AIDS drugs dispute," *San Francisco Chronicle*, November 8, 2001, at A-13, available at http://www.sfgate.com/cgi-bin/article.cgi?file=/c/a/2001/11/08/MN181543.DTL; Paul Blustein, "Drug Patent Dispute Poses Trade Threat: Generics Fight Could Derail WTO Accord," *Washington Post*, October 26, 2001, at E1, available at http://www.washingtonpost.com/ac2/wp-dyn?pagename=article&node=&contentId=A53099-2001Oct25&notFound=true.

and slowed the momentum toward a more trade-friendly world.[56] While such pressure did not mean that the United States would cave to unrealistic demands, it did make an agreement more of a priority than it otherwise might have been.

These external pressures had a definite impact on the negotiations, placing developing nations in a more favorable position than normal. Certainly, developing nations also felt some pressure to compromise. If the Public Health Declaration had not been issued, the status quo would have remained in place, and developing countries would continue to be unsure of their rights when confronting public health issues. Moreover, developing countries did not want to anger the developed countries they rely on for aid money. However, that pressure is present in nearly all negotiations with developed countries. What was different in this instance was that extraordinary events combined to put pressures on developed countries (like the United States) that such countries were not used to feeling and to which there was no evident countermeasure within their political and economic strength.

7.5 Developing Countries' Preparations for Doha

In order to capitalize on these circumstances, developing countries would have to negotiate intelligently. A large part of being a successful negotiator is strong preparation. The developing country coalition exhibited just such a characteristic in the months preceding Doha. Developing countries submitted several official written proposals and were clearly abreast of the important concepts behind the major issues that required resolution. To a certain extent, the papers produced by the developing nation coalition shaped many of the official discussions.

The level of preparation shown by the developed countries was evident as early as the TRIPS Council's first special session dedicated to access to medicines, which took place on June 20, 2001. At this meeting, the Africa Group, side-by-side with many other developing countries, issued a paper stating how it viewed the relationship between the TRIPS Agreement and public health.[57] This paper also

---

[56] *See, e.g.,* "New Hope for WTO Trade Talks" *BBC*, September 3, 2001, available at http://news.bbc.co.uk/1/hi/business/1523321.stm ("The WTO tried and failed to launch a fresh round of trade talks in 1999 at a meeting in Seattle…The WTO is now desperate to iron out disagreements and allow the new round of talks to be launched in November. ") *See* Robert B. Zoellick, "USTR Zoellick Says World Has Chosen Path of Hope, Openness, Development And Growth" *supra* **note 45** ("And we now are delighted that we've overcome the stain of Seattle."); *See* Richard Feinberg, "Why Doha Will do What Seattle Did Not?" *The Straits Times*, November 7, 2001, available at http://www-irps.ucsd.edu/irps/innews/strait110701.html ("Moreover, a second failure would raise questions about the very future of the global trading system and its overseer, the WTO.")

[57] "TRIPS and Public Health," (Submission by the Africa Group, Barbados, Bolivia, Brazil, Dominican Republic, Ecuador, Honduras, India, Indonesia, Jamaica, Pakistan, Paraguay, Philippines, Peru, Sri Lanka, Thailand and Venezuela), WTO, IP/C/W/296, June 19, 2001, available at http://www.wto.org/english/tratop_e/trips_e/paper_develop_w296_e.htm [hereinafter "June 19 Developing Country Group Paper"].

proposed a set of limiting principles on the procedural aspects of the negotiation to come.

The paper consistently cited applicable portions of the TRIPS Agreement and offered valid interpretations of them. For instance, it suggested that Article 6 of the agreement allowed a country to develop its own patent exhaustion regime.[58] Such a reading seems acceptable; Article 6 makes it impossible for a country to challenge another country's patent exhaustion regime through a dispute settlement board.[59] The developing countries also argued that Article 31 could be read as providing for no substantive limitations on the grounds according to which a compulsory license could be issued.[60] Indeed, Article 31 appears to simply set forth a *procedure* that must be followed when a government issues a compulsory license.[61] By maintaining a strong understanding of the legal foundations of the negotiation, the developing countries were able to make logical arguments in support of their interpretation of the agreement.

Of equal importance was that the developing countries attempted to organize the structure of the negotiations. For instance, they noted that the initial discussion on access to medicines should not be viewed as the end of the discussions on this matter,[62] but rather that the negotiations should be ultimately resolved at the Doha Ministerial Conference.[63] They thus placed other countries on notice as to a potential timeframe for the negotiations. Moreover, they emphasized that the focus of the debate should be the link between the TRIPS Agreement and public health.[64] By attempting to limit the subject matter to be discussed, the developing countries were seeking to frame the debate so that they could be sure their interests were addressed. They were also making efforts to prevent the negotiations from becoming a disorganized affair where too many issues were discussed at once and where the parties could be distracted by tangential matters.

In the actual meeting of the TRIPS Council, the Africa Group and its supporters further emphasized both their understanding of the TRIPS Agreement and the negotiating process. Developing countries constantly made reference to the paper and to the points made therein.[65] These countries had clearly come to the meeting

---

[58] *Id.* at paragraphs 24–27.

[59] *Agreement on Trade-Related Aspects of Intellectual Property Rights*, Annex 1C of the Uruguay Round Agreements, at Article 6, available at http://www.wto.org/english/docs_e/legal_e/27-trips.pdf. [hereinafter "TRIPS Agreement"].

[60] June 19 Developing Country Group Paper, at paragraph 32.

[61] TRIPS Agreement, at Article 31. *See also* June 19 Developing Country Group Paper, at paragraph 32.

[62] June 19 Developing Country Group Paper, at paragraph 3.

[63] *Id.*, at Summary.

[64] *Id.* at paragraph 2.

[65] "Special Discussion on Intellectual Property and Access to Medicines," WTO Council for Trade-Related Aspects of Intellectual Property Rights, IP/C/M/31, July 10, 2001, available through "Documents Online" at http://docsonline.wto.org/?language=1 [hereinafter June 22 Minutes].

with an agenda that they wanted to be met. They had planned out a strategy for the meeting and for the negotiating process.

The results were positive. Developing countries appeared as a powerful force that could not be overlooked. The United States and other more cautious countries could not blindly pursue their own agenda without taking notice. Developing countries had not complained of the status quo without presenting a viable solution. Nor had they presented too many solutions that would require an enormous output of time to analyze. To the contrary, they set forth an interpretation of the TRIPS Agreement that was well based in fact. Furthermore, they had laid the groundwork for a process through which their interpretation of the TRIPS Agreement could be analyzed and possibly ratified as being the correct interpretation.

The developing nations' strong preparatory efforts were echoed during the later stages of the negotiations. The informal meeting of the TRIPS Council in July 2001 included yet another contribution from developing countries; the Africa Group issued a statement "[i]n the interest of maintaining the momentum of [the previous] important discussions."[66] In this way, the Africa Group continued to emphasize the course of action they had espoused in June. They offered clarifications of their position with respect to substantive issues like compulsory licensing and parallel importation, and also made a procedural suggestion about the way in which these issues could be fully resolved.[67]

The publication (or the leaking) of this statement was a clever move on the part of the developing nations and NGOs. The July 2001 informal meeting did not leave a formal public record,[68] this document serves as one of the few points of reference for what was discussed. When only limited information is available to the public, it can be advantageous to control what reaches the public's eyes. Furthermore, the statement gave good publicity to the developing country's request, making it appear as if the developing countries were taking a leadership role in the negotiating process. The statement claimed the developing countries wanted to "maintain[] the momentum,"[69] thus implying that they were directing the course of the process.

---

[66] "Statement by the Africa Group," TRIPS and Public Health Informal Session of the WTO TRIPS Council," July 25, 2001, available at http://www.aprnet.org/archive/issues&concern/is01-6.htm [hereinafter "July 25 Statement"]. Other developing countries at the meeting expressed strong support for this statement, although it is unclear whether this statement can be attributed to countries outside of the Africa Group. *See* Cecilia Oh, "TWN Report on TRIPS and Health Session in WTO on July 25, 2001," *Third World Network*, July 27, 2001, available at http://www.aprnet.org/archive/issues&concern/is01-7.htm.

[67] *See* July 25 Statement.

[68] To the best of this author's knowledge, the WTO did not make public the official statements of the trade delegations issued at this informal meeting.

[69] July 25 Statement.

The paper also reaffirmed developing country beliefs with respect to important issues such as compulsory licenses and parallel importation.[70] In so doing, the developing countries solidified their position on substantive matters. Although such a move lessened their flexibility in terms of what they could accept, in this instance it was an intelligent move. It acted as an indication of the seriousness of their intentions, and the logic and rationale of their positions made it difficult to envision an acceptable solution that did not meet at least some of their interests.

The decision by the Africa Group to continue to seek closure on the matter at the Doha Ministerial Conference had an important effect. The Africa Group made it clear that it was unwilling to see this matter put off ad infinitum.[71] This put pressure on countries to resolve the negotiation in a timely manner.[72] The developing countries were thus continuing in their effort to prevent the negotiations from turning into a long episode of discussions that led nowhere.[73]

Another important characteristic of the July 25 Statement was that it put forward in an organized manner a list of requests for what the ultimate declaration should include.[74] Making a request is different from putting forward an interpretation of the TRIPS Agreement. The latter action only ensures that the perspective of the interpreter will be heard, while a request involves asking other parties to judge the validity of this perspective. Of course, it would have been impossible for the developing countries to make a request without having already elaborated a logical reading of the TRIPS Agreement. However, if developing countries had not made a formal request, it would have been unlikely, if not impossible, that aspects of their interpretation of the TRIPS Agreement would have become authoritative.

The Africa Group made known what it wanted out of any resolution of the TRIPS public health issue. Notably, it asked for a statement indicating that nothing in the TRIPS Agreement could prevent a member state from pursuing a public health objective.[75] It also desired a moratorium on dispute settlement; an

---

[70] *Id.* at "Compulsory License" and "Parallel Import" sections.

[71] *Id.*, at third paragraph ("However, we take note of some questions raised by a number of delegations, and we look forward to reaching agreement on those points, so that we may proceed swiftly towards achieving a tangible result for our discussions, in time for Doha."), and at concluding paragraph ("We feel it is time that all delegations set out in specific terms what they would want to see done in the period leading up to Doha, and at Doha, regarding the TRIPS Agreement and its impact on access to affordable medicines.")

[72] *Id,* at concluding paragraph.

[73] One could argue that developing countries failed in this respect. Trade delegations from around the world are still trying to reconcile TRIPS and public health to this date (Summer 2003). Nevertheless, it is a credit to the developing nations that they could at least achieve the Public Health Declaration in November 2001.

[74] July 25 Statement, at "Preparatory Work for Doha."

[75] *Id.*

extension of the transition period for Least Developed Countries (LDCs) and developing countries to implement the TRIPS Agreement; and recognition of the preeminent importance of Articles 7 and 8 for the interpretation of the TRIPS Agreement.[76]

These requests were derived from developing countries' earlier arguments. For instance, in their first paper the developing countries had set forth the notion that "nothing in the TRIPS Agreement will prevent Members from adopting measures to protect public health."[77] Elsewhere in that paper, the developing countries had declared that "each provision of the TRIPS Agreement should be read in light of the objectives and principles set forth in Articles 7 and 8."[78] Lastly, the requests for a moratorium on dispute settlement and extension of the transition period logically follow from the Africa Group's statement during the June 22 meeting.[79] Thus, the Africa Group managed to skillfully use the arguments and requests set forth prior to the July 25 meeting in order to develop a list of requests that made sense within the context of the negotiations. Other countries would now have to react to their proposals.

By September, the Africa Group and its allies had gone one step further. They produced a complete draft declaration for consideration in the preparatory phases of the Doha meeting.[80] This draft again offered interpretations of how compulsory licensing could be pursued, and it sought to affirm that public health took precedence over intellectual property rights.[81] However, it also sought to establish other means by which governments could provide for access to medicines. For instance, one proposal in the draft would have allowed generic pharmaceutical manufacturers, without the permission of the patent holder, to produce patented medicines for export in order to address the public health concerns of the importing countries.[82] There was also a suggestion that the compulsory licensing procedure could be bypassed if it was found that the patent holder was engaging in "anti-competitive behavior."[83]

---

[76] *Id.*

[77] June 19 Developing Country Paper, at paragraph 22.

[78] *Id.*, at paragraph 17.

[79] June 22 Minutes, at 6.

[80] "Draft Ministerial Declaration" (Proposal by the African Group, Bangladesh, Barbados, Bolivia, Brazil, Cuba, the Dominican Republic, Ecuador, Haiti, Honduras, India, Indonesia, Jamaica, Pakistan, Paraguay, the Philippines, Peru, Sri Lanka, Thailand and Venezuela), October 4, 2001 (document originally distributed September 2001 special meeting of the TRIPS Council), IP/C/W/312, available at http://www.wto.org/english/tratop_e/trips_e/mindecdraft_w312_e.htm. [hereinafter "September Draft Ministerial Declaration"].

[81] *Id.*, at paragraphs 1, 4.

[82] *Id.*, at paragraph 9. Such a proposal was presented as an Article 30 exception. *See id.* Article 30 of the TRIPS Agreement allows for "limited exceptions to the exclusive rights conferred by a patent."

[83] September Draft Ministerial Declaration, paragraph 6.

Introducing these more sweeping proposals served a purpose in the negotiations. It was unlikely that these or similar proposals would become part of the final declaration.[84] However, they made developing countries' primary requests appear far more realistic. In addition, it would have been senseless for developing countries to settle for one result when they could have achieved greater concessions which they believed would have aided their cause even further. Finally, making such requests may have unsettled their negotiating counterparts. Were these countries actually trying to undermine the integrity of the TRIPS Agreement? Such a result, although certainly unattainable, must have been worrisome to Western trade ministers in an environment in which the TRIPS Agreement was increasingly coming under attack.

7.6 Developed Countries' Preparations for Doha

Throughout this period, developed countries like the United States were not just ignoring the issue, nor did they fail to understand it properly. However, the pharmaceutical-producing nations were in a different position. They did not feel the same degree of urgency to make any changes to the TRIPS Agreement.[85] They therefore adopted more of a "wait-and-see" approach as the TRIPS Council meetings on access to medicines began. The United States did not submit a paper on the issue during the first June meeting of the TRIPS Council. In their statement to the Council, the United States expressed an interest in limiting the solutions that could be entertained, while seeking to promote public health. For instance, they confirmed that the TRIPS Agreement "has struck a proper balance between offering incentives for innovation and ensuring that there is access to needed medicines."[86] They thereby signaled an unwillingness to modify the TRIPS Agreement itself. Furthermore, the U.S. delegate presented cautionary words about the use of compulsory licenses.[87] Finally, the American delegate noted that while

---

[84] Switzerland announced in its statement at the September meeting that it was "concerned about any possible weakening of intellectual property protection in [the public health] field." TRIPS Council, "Minutes of Meeting Held in Centre William Rappard on 19 and 20 September 2001," IP/C/M/33, November 2, 2001, paragraph 178, available through http://docsonline.wto.org/gen_search.asp?searchmode=simple. It is to be presumed that the United States felt similarly. Nevertheless, the developing country coalition had proposed solutions that provided for broad changes to the intellectual property regimes in practice. On the one hand, they had proposed revising the text of the TRIPS Agreement itself (the "anti-competitive practices" exception). On the other hand, they had proposed a more global public health exception to traditional patent rights. In light of the posture of the developed country coalition, it would have been very surprising if either of these proposals had achieved success. Indeed, the final Public Health Declaration bears no reference to either of these proposals. .

[85] *See supra* notes 50, 83.

[86] June 22 Minutes, at 36.

[87] *Id.* at 37–40.

Article 6 of the TRIPS Agreement prevented governments from introducing actions against another government for parallel importation of a product, the Article did not explicitly authorize parallel importation.[88] The general tenor of the American statement was one of reluctance to modify or challenge any prior understanding of the way the TRIPS Agreement functioned.

The developed countries' initial strategy was clear: Be prepared to discuss the issues, but hesitate before taking any steps that might limit the scope of the TRIPS Agreement. Thus the developed countries were not exactly presenting a road-block,[89] but were rather slowing the negotiation process so that no decisions were made in haste. Due to the atmosphere of the time, the United States could not afford to actively combat the mission of the developing countries.[90] However, they also did not want to make decisions that might prove unnecessarily harmful to their interests.

As the time came to prepare for the Ministerial Conference at Doha, the United States and other developed countries had to give some formal indication of their thinking as to an appropriate solution. To do this, the United States and four other countries introduced draft language for a preamble to the declaration.[91] Their draft preamble text tended to limit the focus of the declaration to pandemics like AIDS rather than public health generally.[92] Furthermore, it sought to reaffirm that the TRIPS Agreement "contributes to the availability of medicines" and that WTO Members remain "commit[ed] to the TRIPS Agreement and its implementation."[93] The preamble gave little sense that the five countries supporting it were willing to take more than a piecemeal approach to addressing the relationship of TRIPS and public health.[94] The developed countries clearly wanted to avoid unnecessarily broad language that could have unforeseen and unpredictable effects.

---

[88] *Id.* at 40.

[89] For instance, the American delegate noted at the June 22 meeting that it "look[ed] forward to continuing [the public health] discussion in future meetings and to all [WTO] Members' contributions." *Id.* at 40.

[90] Thus, the American delegate noted that "[t]he HIV/AIDS crisis is a terrible tragedy–for countries, families and individuals. The United States is fully committed to the battle against this disease." *Id.* at 33.

[91] "Preambular language for ministerial declaration" (Contribution from Australia, Canada, Japan, Switzerland and the United States), IP/C/W/313, October 4, 2001, available at http://www.wto.org/english/tratop_e/trips_e/mindecdraft_w313_e.htm [hereinafter "September Draft Preamble"].

[92] For instance, the subtitle of the September Draft Preamble was "access to medicines for HIV/AIDS and other pandemics." *Id.* Moreover, the WTO Internet site itself notes in its summary of the September Draft Preamble that the focus is "more closely on tackling problems such as pandemics (HIV/AIDS, malaria, TB)." "Members discuss drafts for ministerial declaration," WTO News: 2001 News Items, TRIPS Council Wednesday 19 (and 21) September 2001, available at http://www.wto.org/english/news_e/news01_e/trips_drugs_010919_e.htm.

[93] September Draft Preamble.

[94] *See id.*

This positioning was not surprising. Developed countries characteristically have greater power in negotiations due to their economic might, and they can afford to enter into many costly dispute resolution procedures simultaneously. As noted earlier, developed countries also hold the key to unlocking vast amounts of financial aid to impoverished countries. Thus, developed countries can often get what they want because they control factors that are external to the negotiations but that have a great impact on the decision-making of developing countries. In light of this, developed countries often will not be in a hurry to agree to provisions that change the status quo or that will have an unknown effect on the status quo.[95]

7.7 WTO General Council Involvement

The initial stages of the TRIPS public health negotiations thus were characterized by high levels of preparation by all participants. In effect, both sides established strategies to safeguard their aims and followed through on these strategies.

At the end of the September special TRIPS Council Meeting, however, some distance between the parties clearly remained. The developing countries wanted a clear statement, along with proof, that the TRIPS Agreement would not keep them from reaching their public health objectives. At the same time, the United States and other developed countries wanted to make sure that intellectual property rights were not undermined. Throughout October and indeed until the Doha Ministerial came to a close, the major question was whether it was possible to reconcile these competing interests.

In October 2001 the negotiations shifted away from the TRIPS Council and toward the WTO General Council. Because the Public Health Declaration was going to be issued at the WTO Ministerial Conference, only the General Council had the authority to issue it. General Council Chairman Stuart Harbison, aided by the Director-General of the WTO Secretariat Michael Moore, was charged with preparing draft declarations for discussion at the Doha Ministerial Conference.

The production of drafts by Chairman Harbison was critical to the negotiating process. It was unlikely that any of the negotiating parties would agree to work from a draft produced by their counterparts. Theoretically, Harbison could act as an honest broker between the two negotiating coalitions and produce a draft that reflected the interests of each side.[96]

---

[95] For reasons why developed countries lost much of their leverage in the context of the public health negotiations, see *supra* pages 11–20.

[96] Some participants in the negotiating process did not feel that Chairman Harbison pursued the appropriate methodology when producing the draft language. *See* Cecilia Oh, "Draft Doha texts fail to reflect developing country concerns on TRIPS," *The Road to Doha: Divisions and Dissent on New WTO Round*, Third World Network, available at http://www.twnside.org.sg/title/twr133f.htm.

The texts that Harbison issued are also a good way to measure the progress each side had made in the prior months and the amount of negotiating that remained. Harbison's first draft text was produced on October 21, 2001.[97] The "Draft Elements," as they were called, were not a draft Declaration but rather an indication of what future drafts might contain. Developing countries could note with some satisfaction that the Draft Elements included recognition of their right to resort to some form of parallel importation or compulsory licensing.[98] However the U.S.-led coalition had to be even more pleased, since the Draft Elements were by and large very conservative. Most importantly, the Draft Elements included only a tepid version of the developing country request for recognition that public health concerns trumped intellectual property rights. The Draft Elements "emphasize" that the TRIPS Agreement "permits" a government to take measures to "[protect public health] [to secure access to medicines at affordable prices]."[99] Such a statement really added nothing of substance to the negotiations. No country would have agreed to the TRIPS Agreement in the first place if it had not permitted countries to seek ways to promote public health. The developing country concern was that the TRIPS Agreement might be interpreted to *limit* a government's options when choosing a means to protect public health.

Just 1 week later, developing countries regained a foothold in the negotiation. The new October 27 Draft[100] included language with respect to the relationship between the TRIPS Agreement and public health that was much more favorable to the developing countries.[101] This language was only presented as one option from which to choose[102]; however, it fairly represented what these countries were seeking.

It is unclear what caused this sudden about-face in the draft language. However, it can most likely be attributed to two factors. First, developing countries and NGOs must have made their displeasure with the Draft Elements known to Harbison.[103] Second, the developing countries had extensive documentation of their true position; it would have been difficult for Harbison or anyone else to claim that he or she was ignorant of their argument.

---

[97] "Elements for Draft Declaration on Intellectual Property and [Access to Medicines] [Public Health]," October 21, 2001, available at http://www.ictsd.org/ministerial/doha/draftTRIPS21Oct.pdf [hereinafter "Draft Elements"].

[98] *Id.* at paragraph 5.

[99] *Id.* at paragraph 4.

[100] "Draft Declaration on Intellectual Property and [Access to Medicines][Public Health]," (as prepared by the Chairman of the General Council, in cooperation with the Director General [of the WTO]), October 27, 2001, available at http://www.ictsd.org/ministerial/doha/docs/IP27oct.pdf. [hereinafter "October 27 Draft"].

[101] *Id.* at paragraph 4.

[102] *Id.*

[103] For an example of disappointment in the Draft Elements, see Cecilia Oh, "An Update on the Progress of Consultations on the Proposed Ministerial Declaration on TRIPS and Public Health," available at http://lists.essential.org/pipermail/ip-health/2001-October/002266.html.

In any event, the October 27 Draft shows that the negotiation was still to be completed. While the developing countries might have been pleased to see their proposal reflected in this draft, the developed countries' competing vision was also presented as a choice.[104] This latter proposal again failed to say that TRIPS obligations were secondary to public health concerns.[105] To the contrary, it stated that the Public Health Declaration "does not add to or diminish the rights and obligations of members provided in the TRIPS Agreement."[106] This paragraph would remain a source of controversy until the ministerial meetings themselves drew to a close.

With respect to the other issues, the October 27 Draft was also more favorable than the Draft Elements to the developing countries' interests. For instance, it explicitly affirmed the right to "grant compulsory licenses and the freedom to determine the grounds upon which such licenses are granted."[107] On the other hand, certain provisions suggested by the U.S.-led coalition were also included in the October 27 Draft, most notably the transition period for LDCs to implement the TRIPS Agreement with respect to "pharmaceutical products" and the period during which there would be a moratorium on dispute settlement.[108] As Cecilia Oh from the NGO Third World Network noted, the concern with the transition period or moratorium was mainly that the United States was attempting to "[entice] the LDCs and African countries into accepting the proposal in exchange for a weakened declaration, and [to break] up the developing country coalition."[109] With respect to the major concepts that had been discussed during the negotiations, however, the October 27 Draft contained elements of those that had most preoccupied the developing countries.

The WTO draft declarations thus revealed that the developing countries had made great gains over the course of the months leading up to the Doha Ministerial meetings. Many of their negotiating perspectives were accepted without much controversy. However, the largest issue was still unresolved. The ultimate relationship between TRIPS and public health was of great importance to all of the participants in the negotiation. Nevertheless, the developing countries were being heard. It was just a matter of whether a solution could be reached whereby all interests could be met.

---

[104] October 27 Draft, at paragraph 4.

[105] *Id.*

[106] *Id.*

[107] *Id.* at paragraph 6.

[108] *Id.* at paragraphs 10, 11. *See also* Oh, "Draft Doha texts fail to reflect developing country concerns on TRIPS," *supra* note 96.

[109] Oh, "Draft Doha texts fail to reflect developing country concerns on TRIPS," *supra* note 96.

7.8 Doha/Conclusion

Such a solution was indeed reached. By November 12—the middle of the Doha Ministerial—the Draft Declaration[110] was strikingly similar to the Final Public Health Declaration. The only major modifications were the decision to use "can and should" instead of "shall" in the paragraph describing the relationship between the TRIPS Agreement and public health, and the removal of the moratorium on dispute settlement from the Public Health Declaration.[111] By November 12, participants had decided upon general statements that the "TRIPS Agreement does not and should not prevent Members from taking measures to protect public health" and, that the TRIPS "Agreement [can and should] [shall] be interpreted and implemented in a manner supportive of WTO Members' right to protect public health and, in particular, to ensure access to medicines for all."[112] At the same time, the parties agreed to "reiterate [their] commitment to the TRIPS Agreement."[113] Basically, the participants had agreed to incorporate the interests of both sides as closely as possible into the same statement.

These interests were not easily combined, and it is striking how difficult it is to paraphrase the final version of this paragraph. Almost any language other than that found in the Public Health Declaration appears to distort the meaning of the paragraph.

How the final result was attained is difficult to determine. The negotiating at Doha was for the most part conducted outside of the public eye. Therefore we do not know what negotiating tactics or strategies were employed to reach the final result. That is not to say that the negotiations should have been more transparent. People need privacy to brainstorm and develop solutions. Without such privacy, negotiators may be embarrassed at the public exposure given to a failed proposal or a potential solution that has a glaring error.

What then can be said about the public health negotiations that took place in Doha? Clearly, these negotiations took place by and large among a limited number of negotiators who represented the various factions.[114] They worked from a draft text and had to resolve the final language difficulties. Meanwhile, it is almost certain that other negotiators were discussing the issue on a more informal basis outside the well-known "green room" to look for potential compromises.

---

[110] "Draft Declaration on the TRIPS Agreement and Public Health," November 12, 2001, available at http://www.ictsd.org/ministerial/doha/docs/IP12nov.pdf [hereinafter "November 12 Draft"].

[111] *See id.* at paragraphs 4, 8; Public Health Declaration, paragraph 4, available at http://www.wto.org/english/thewto_e/minist_e/min01_e/mindecl_trips_e.htm.

[112] November 12 Draft, at paragraph 4.

[113] *Id.*

[114] The major official participants at Doha were the United States, Brazil, Nigeria, Kenya, the European Union, New Zealand, Zimbabwe, India, Peru and Switzerland. *See* Godel, *supra* note 35, at 52.

Within the more formal setting, the WTO had appointed "friends of the chair" for each of the key issues discussed. These individuals were charged with facilitating the negotiations among nations and were supposed to be entirely neutral. The other negotiators most likely employed standard negotiating tactics. Certainly, they made proposals. Certainly, they argued and haggled over ideas. They also tried to pressure their counterparts to accept various solutions.

Whose tactics worked best? This question is impossible to answer. The fact is that both sides got much of what they wanted. The Americans and other developed nations could claim that the Doha Ministerial was a success and could accept the Public Health Declaration. The developing nations could also point to the Public Health Declaration as a success.[115] In truth, there were no real losers in the negotiation, aside from perhaps the pharmaceutical industry.

The question of whether there were actually any winners can of course only be answered in the future. If the AIDS epidemic is controlled in the coming years, then the Public Health Declaration can be hailed as a positive step. If the AIDS epidemic continues to rage unchecked, it will be difficult to say that the Public Health Declaration had much impact. Countries would then just have to settle for the notion that the Public Health Declaration represented an attempt by the world to address this health crisis.[116]

That leads to the ultimate question. Why did this Declaration come into existence?[117] To a certain extent, the Public Health Declaration came about because of the talent of the negotiators and the dedication of the various NGOs. The preparation of each side also allowed for the development of a Declaration that squarely addressed many of the important issues. The use of coalitions helped to ensure that developing countries were able to influence the direction of these negotiations, and exterior factors like the anthrax crisis produced the pressure to come to agreement. On a more global level, the Declaration came about for one major reason: it was the right thing to do at the time.

---

[115] For instance, many NGOs claimed the Public Health Declaration as a success. *See supra* note 9.

[116] Of course, as history has shown, the Public Health Declaration did not result in a quick resolution of all of the outstanding issues. Even as of this writing, there is still a major international trade controversy raging about aspects of the relationship between TRIPS and access to medicines. .

[117] An article in the *Journal of the American Medical Association* stated that the Public Health Declaration came about because of (a) NGO influence; (b) the formation of coalitions; and (c) the controversy surrounding the Cipro patent during the anthrax crisis. *See* David Banta, "Public Health Triumphs at WTO Conference", JAMA. 2001; 286 (21) 2655–2656.

Chapter 8
Case II—Negotiating Access to HIV/AIDS Medicines: A Study of the Strategies Adopted by Brazil

Abstract The bilateral dispute between Brazil and the United States with regard to Brazil's protection of intellectual property has gone on for longer than a decade without a definite resolution. The issue gained momentum when Brazil introduced a program of fighting AIDS and made changes in domestic legislation to facilitate its implementation. The U.S. believes that Brazil's actions were in direct violation of its obligations to protect intellectual property rights (IPRs) under the multilateral agreement on Trade-Related Aspects of Intellectual Property Rights (TRIPS). Brazil maintains that it has full legitimacy to use all necessary means to save its population from the AIDS pandemic.

Keywords HIV/AIDS · Brazil · United States of America (USA) · United States Trade Representative (USTR) · Issue framing · Negotiation process · World Trade Organization (WTO) · Pharmaceuticals · TRIPS · PhRMA · NGOs · Intellectual property rights · Coalition building · Compulsory licensing · International consensus-building · Generic drugs · Access to essential medicines · BATNA

Although the dispute has not been resolved, Brazil, while clearly a weaker party than the United States, has been very successful in its negotiations with the world's superpower. Ultimately, the U.S. withdrew its formal complaint from the World Trade Organization (WTO) over Brazil's domestic legislation, and the two parties decided to conduct negotiations in a bilateral forum.[1] This action had the effect of

---

Case study researched and written by Anand Balachandra and Mariya Kravkova.

---

[1] "US beats a (tactical) retreat over Brazil's patent law," by Charkravarthi Raghavan, "Third World Network," http://www.twnside.org.sg/title/tactical.htm.

D. Fairman et al., *Negotiating Public Health in a Globalized World*,
SpringerBriefs in Public Health, DOI: 10.1007/978-94-007-2780-9_8,
© The Author(s) 2012

easing pressure on Brazil and allowing the Brazilian government more flexibility in pursuing its health policy, while also saving the Bush administration from a public relations disaster at home and abroad.

This case sheds light on strategies that a weaker party (a "lamb") can use to strengthen its case vis-à-vis a more powerful player (a "lion"). The first section introduces the dispute, the main stakeholders, the issues and the interests of the parties. The second section takes a close look at the partnership between the Pharmaceutical Research and Manufacturers Organization (PhRMA) and the United States Trade Representative (USTR) and discusses the implications for Brazil of the benefits and costs of this partnership. The third section analyzes the negotiation strategy of a "lamb," focusing on Brazil's strategies and tactics. The fourth section analyzes the current state of affairs, examines the implications of pursuing this dispute in a bilateral forum and the lessons that should be learned from the negotiation experience, and suggests some strategies for the future. The paper concludes that Brazil should develop new tactics to augment its power under the changed circumstances.

8.1 Overview of the Dispute

The origins of this dispute go back to a 1987 petition filed with the USTR by the Pharmaceutical Manufacturers Association (PMA), a predecessor of PhRMA.[2] The PMA alleged that Brazil lacked appropriate patent protection laws and that the situation was detrimental to U.S. commercial interests. Following the petition and a recommendation by the USTR, President Reagan increased tariffs on several categories of goods from Brazil, pursuant to Section 301 of the Trade Relations Act of 1974.[3] The dispute escalated further in 1993, when the USTR recognized Brazil as a Priority Foreign Country based on Special 301 provisions of the 1974 Act.[4]

In response to this trade pressure, the Brazilian legislature passed several laws providing for stronger monopolies for pharmaceutical patents. However, the legislation fell short of what the USTR wanted. As the two countries were unable to resolve their differences through negotiation, the U.S. on April 30, 2000, requested that the WTO establish a dispute resolution panel to review Brazil's patent law. The United States' main objection was a local manufacturing requirement found in Article 68 of Brazil's law. Article 68(1)(I) stipulated that if a patented product is not being manufactured in Brazil within 3 years from the

---

[2] Consumer Project on Technology, Bilateral Trade Disputes involving the United States, over intellectual property and health care, September 2000.

[3] Clyde Farnsworth, "Reagan Imposes Punitive Tariffs Against Brazil" The New York Times, November 14, 1987. http://www.nytimes.com/1987/11/14/business/reagan-imposes-punitive-tariffs-against-brazil.html

[4] Report to Congress on Section 301 Developments Required by Section 309(a)(3) of the Trade Act of 1974, January 1995–1996.

date of issuance of the patent, the government may "compulsory license" a competitor. The U.S. views this requirement as inimical to free trade and a violation of the TRIPS agreement.[5]

8.2 The Main Stakeholders and Their Interests

The debate over intellectual property and access to essential medicines has many stakeholders. Among them are the governments of the two countries, U.S. pharmaceutical companies, Brazil's pharmaceutical industry and of course AIDS patients. This case focuses on the three major players in the negotiation process—the Brazilian government, the U.S. government (represented by the USTR) and PhRMA. What follows is a discussion of each of these stakeholders and their respective interests.

8.2.1 Brazil

The Brazilian government and the Brazilian people are among the principle stakeholders in this negotiation. Brazil has great incentive to obtain drugs from the pharmaceutical companies. The commonly used "triple cocktail" therapy still does not cure AIDS (it makes the disease chronic), so entirely new drugs must be developed to cure the illness.[6] Due to the high cost of developing new drugs, Brazil is dependent on the West to find an ultimate cure. This has not prevented the Brazilian government, however, from contemporaneously producing generic versions of currently available drugs locally, to lower prices and make the therapy more accessible to Brazilian AIDS patients. Local production is a crucial element of Brazil's national health policy, because it allows the government to access drugs at affordable prices.

During the negotiation process, Brazil consistently attempted to define the issue as that of "access to essential medicine." In the process, the Brazilian government sought to broaden the focus of the negotiation from the protection of intellectual property (IP) rights to include additional economic, security, social and public health concerns. In the past, pure price negotiations have resulted in only limited victories for the governments of developing countries. (In Senegal, Rwanda and Uganda, for example, "successful" negotiations led to only a very small number of affected persons receiving the needed drugs, due in part to a lack of supporting distribution systems).[7] To achieve greater results, Brazil took a "multi-sector"

---

[5] Office of the United States Trade Representative, Executive Order of the President, Washington, DC, June 25, 2001, at 2. http://www.ustr.gov/releases/2001/06/01-46.htm.

[6] See Rochelle Jones, Scientists Discover a Cureall in Treating HIV Infection (Nov. 11, 1999) http://www.cnn.com/HEALTH/AIDS/9911/11hiv.hide.journal.

[7] A supporting distribution system is crucial in the case of AIDS treatment since the ultimate success of therapy depends, to a large degree, on its correct application.

approach to the negotiations and included agents from international lending organizations, nongovernmental organizations (NGOs), the religious community and the local business community. This approach enabled Brazil to move the issue from purely "price negotiations" to one dealing with funding for the acquisition and distribution of drugs, as well as toward broader concerns of how a country may cope with national emergencies under current international IP law.

8.2.2 The USTR and the Pharmaceutical Companies

The U.S. government and the pharmaceutical companies argue that what is at issue is innovation and implications for international trade, as well as the industry's ability to finance research and show a profit.[8] As private players relying on equity financing, U.S. pharmaceutical companies without doubt are also concerned about maximizing shareholder value. However, PhRMA has not framed the issue as purely one of profit maximization. The pharmaceutical companies argue that strong patent rights are crucial to the survival and success of the industry. If companies are not able to achieve a certain level of profits, they are unable to undertake research and development of new drugs. Shannon Herzfeld of PhRMA has argued that pharmaceuticals are very dependent on the rules-based system that preserves strong IPRs, because "out of the 15,000 molecules tested, only three are suitable for human use and only one becomes profitable. The cost of this research amounts to $500 million and takes 12 years to develop."[9] That is why a period of market exclusivity is needed. Further, Ms. Herzfeld noted, "this debate does not represent a North versus South issue. No one has a monopoly on good ideas. Without a rules-based system [providing strong protection of IPRs], expediency is raised over sustainable development."[10]

The U.S. government looks at the issue from the perspective of IPR protection and enforcement of international agreements. In international trade and trade-related negotiations, the U.S. government's interests are represented by the USTR.[11] The USTR relies heavily on Special Section 301 pursuant to the Trade Act of 1974 to enforce U.S. rights under international trade agreements.[12] Since

---

[8] See *The New York Times*, "Do the Poor Have a Right to Cheap Medicine?" June 25, 2000 Sunday, Section 4, page 18, Column 1; Week in Review Desk.

[9] Personal notes of the author, who was present at the conference.

[10] *Id.*

[11] I will use the term U.S./USTR interchangeably throughout the case, unless otherwise noted.

[12] "Special 301" is Section 182 of the 1974 Act. It was added to the 1974 Act by Section 1303 of the 1988 Act. Special 301 requires the USTR to provide information on an annual basis to Congress about countries that lack or fail to enforce IPRs. The fundamental purpose of Special 301 is to increase the USTR's leverage in negotiations aimed at trade liberalization. Critics say that Special 301 is a "heavy-handed tool that compels America's trading partners to negotiate under duress." See Raj Bhala, *International Trade Law: Theory and Practice*, Second Edition, Lexis Publishing, at 1258.

the PMA submitted its petition to the USTR in 1987, the U.S. government has lobbied the Brazilian government to change its domestic legislation in a manner consistent with U.S. interpretation of international law.

As a result, in April 1996 Brazil enacted "a new, long-awaited industrial property law, providing patent protection and greater market access for products."[13] However, the USTR alleged that Brazil's new patent law (Article 68) violated international trade rules by requiring local production in order for a foreign patent holder to receive protection in Brazil. According to the USTR's position, the "local working" requirement is imposed unfairly on patent holders and discriminates between imported patented products and those made locally. Specifically, they said, it violates TRIPS rule 27.1, which says that patents "shall be available and patent rights enjoyable without discrimination as to the place of invention, the field of technology and whether products are imported or locally produced."[14] The U.S. maintains that Article 68 of Brazil's law discriminates against U.S. owners of Brazilian patents whose products are imported and not produced in Brazil.[15]

The interests of PhRMA and the USTR are aligned to a great degree, except that the USTR's approach is broader. That is, the USTR concerns itself with overall trade policy, while PhRMA's main concern is the preservation of its own commercial interests. Insofar as the USTR's approach recognizes the need to preserve PhRMA's commercial interests as part of overall trade policy, however, the interests of the two parties are shared. This alignment of interests has affected the formulation of PhRMA's strategy.

8.3   PhRMA's Strategy

PhRMA has engaged in a multi-track strategy, but the most important facet of its strategy has been its reliance on the USTR to pursue its goals. This strategy has been quite successful; however, it has not come without costs. Analyzing PhMRA's strategy is beneficial for both sides to the dispute. From Brazil's perspective, it may be valuable to examine carefully the benefits and drawbacks of the partnership, as the particulars of the relationship bear direct relevance on how Brazil should formulate its own negotiating strategy.

---

[13] James Love, Consumer Project on Technology, Interview available on http://www.cptech.org.

[14] Agreement on Trade-Related Aspects of Intellectual Property Rights (TRIPS), reprinted in 2000 *International Trade Law* 567, 579 (Raj Bhala, Lexis Publishing 2000).

[15] World Trade Organization, WT/DS199/3 9 January 2001(01-0093), Brazil—Measure Affecting Patent Protection, Request for the Establishment of a Panel by the United States, found on http://www.cptech.org/ip/health/c/brazil/Req4EstabPanel.html (accessed on November 7, 2001).

8.3.1 Benefits of the USTR/PhRMA Partnership

Although PhRMA continues to negotiate directly with the Brazilian government, U.S. pharmaceutical companies understand that their leverage vis-à-vis a sovereign government is rather small compared to that of the USTR. The USTR has proven to be capable of "convincing" sovereign states to change their domestic laws and policies, while the pharmaceuticals industry's success in direct negotiations with Brazil has been confined to achieving less rigorous price cuts than what the Brazilian government initially wanted. The realization of this fact led PhRMA to rely increasingly on the USTR to achieve its goals.

PhRMA has lobbied and used its multiple government connections to influence USTR policy. Indeed, PhRMA is considered one of the most aggressive trade groups in Washington. It has acquired seats on important advisory boards that shape government policy, including a special presidential advisory group on trade. Moreover, as one commentator has stated, "prominent government officials [have] spun through the revolving door between government and the pharmaceutical industry."[16] PhRMA's top representatives have included, among others, Gerald Mossinghoff, previously Assistant Commerce Secretary and Commissioner of Patents and Trademarks during the Reagan administration (1981–1985), and Harvey E. Bale, Jr., previously a senior USTR official. In one interview, Mr. Bale conceded that his government connections were a plus for the industry: "I don't hide it. In fact, I'm happy to help them out."[17]

Observers note that the industry's success in lobbying and its effective use of government connections manifested itself in the following ways. First, the Office of the USTR exerted "extraordinary pressure" on individual developing countries to adopt U.S.-style patent laws.[18] Second, the USTR insisted that intellectual property provisions be included in the Uruguay Round negotiations of the General Agreement on Tariffs and Trade (GATT), which were completed in 1994.[19] Third, provisions for IPR protection were included in the North American Free Trade Agreement (NAFTA) as a central component, as well as considered by the U.S.

---

[16] Robert Weissman "A Long, Strange TRIPS: The Pharmaceutical Industry Drive to Harmonize Global Intellectual Property Rules, and the Remaining WTO Legal Alternatives Available to Third World Countries." 17 U.Pa. J. Int'l Econ. L. 1069 (1998), 1076.

[17] See Julie Kosterlitz, Rx: Higher Prices, NAT'L J., Feb. 13, 1993, at 77.

[18] U.S. Trade Representative Clayton K. Yeutter recalls that when he left government service at the end of the Ford administration, hardly anyone in Washington had ever heard of the notion of intellectual property. He returned to government two years ago to find it one of the hottest buzzwords in town. "Intellectual property issues have become central to congressional debate on trade policy," says Sen. Patrick Leahy (D-Vt.). Robert Wisseman, "A Long, Strange TRIPS: The Pharmaceutical Industry Drive to Harmonize Global Intellectual Property Rules, and the Remaining WTO Legal Alternatives Available to Third World Countries." 17 U. Pa. J. Int'l Econ. L. 1069; John Burgess, *Fighting Trespassing on "Intellectual Property": U.S. Tries to Prevent Overseas Copying of Everything from Music to Microchips,* WASH. POST, Dec. 6, 1987.

[19] See Burgess, *supra* note 18.

government to be the USTR's top priority in negotiations over the Free Trade Area of the Americas (FTAA).[20]

With regard to Brazil, as has already been mentioned, in 1988 President Reagan used a trade relations clause of the 1974 Trade Act (Section 301) to increase tariffs on incoming goods from Brazil.[21] This was only 1 year after the PMA (now PhRMA), filed a petition with the USTR asserting that Brazil lacked patent protection laws for pharmaceutical products and processes to manufacture them.[22] Since then, the U.S. has been putting pressure on Brazil to change its law. In 1993, for example, the USTR recognized Brazil as a Priority Foreign Country based on Special 301 provisions.[23] This trade pressure resulted in changes to Brazilian domestic laws. Even though the benefits to the pharmaceutical industry were not immediate, the USTR's pressure was quite effective.

Thus, the alliance of PhRMA with the USTR on this issue is quite logical from PhRMA's perspective. From a negotiating standpoint, PhRMA's strategic decision to use the USTR as its agent in international negotiations adds powerful leverage. However, it also carries some drawbacks.

8.3.2 Drawbacks of the USTR/PhRMA Partnership

There are multiple risks embedded in PhRMA's reliance on the USTR to defend its interests. Some of these risks relate to the challenges that all international agents face in their attempts to pursue their principals' interests.[24] (In this case, the USTR is the agent and PhRMA is one of the principals the USTR represents.) These challenges can be looked at as costs incurred by the principal in cases when they diminish the ability of an agent to pursue effectively principal's interests. Such challenges include, but are not limited to, serving multiple principals, shifting mandates and role conflicts. As discussed later in this case, at least one facet of Brazil's strategy should be (and to a degree has been) to capitalize on these agency costs to increase its negotiating power.

The USTR, like most agents, serves multiple principals. One scholar defines the "multiple principals" problem as "the most notable feature of international diplomatic negotiation."[25] The USTR's principals include the U.S. President,

---

[20] See *International Agreements: Intellectual Property Enforcement to Play Major Role in NAFTA Talks,* 8 Int'l Trade Rep. (BNA) No. 42, at 1553 (Oct. 23, 1991).

[21] Tech Law Journal, WTO Upholds Section 301 of Trade Act, January 2000.

[22] Consumer Project on Technology, Bilateral Trade Dispute Involving the United States, over intellectual property and health care. September 2000.

[23] Report to Congress on Section 301 Developments Required By Section 309(a)(3) of the Trade Act of 1974, January 1995–June 1996.

[24] Eileen F. Babbitt, "Challenges for International Diplomatic Agents," in *Negotiating on Behalf of Others*, ed. by Robert H. Mnookin, Lawrence E. Susskind, at 136.

[25] *Id.*

numerous trade lobbies and indirectly, some non-trade lobbies. Unlike in the classic principal-agent model, a number of people are involved in this case who contribute to the differences between the interests of the principal and those of the agent.[26] In addition, there are differences between the multiple principals themselves. Clearly, the interests of the U.S. government are much broader than those of PhRMA. The USTR is responsible for formulating sensible trade policies, while PhRMA's primary goal is to increase shareholder wealth.

Another challenge to the relationship between PhRMA and the USTR is the "changing mandate" problem. The more influential a principal is, the more it is capable of changing the mandate of the agent. For the USTR, the U.S. government is the most influential principal and thus is capable of affecting the USTR's mandate in various ways, not all of which may be favorable to PhRMA. Moreover, the U.S. government is the "elected" principal [27] and therefore is susceptible to the pressures and influences of a broad range of constituents, including those in direct opposition to PhRMA.[28] For example, a public health NGO who lobbies the USTR is likely to focus on access to essential medicines and unlikely to include profit maximization as a goal. Thus, the USTR's mandate may be a "moving target." This is a cost to PhRMA, but presents a special opportunity for Brazil to lobby for a change in U.S. government policy.

The White House executive order of May 2000 presents one example of how the "moving target" mandate of the USTR can directly affect PhRMA's interests in a negative way. The executive order stated that the U.S. government would not oppose African nations that violated American patent law to get AIDS drugs.[29] This was such a tremendous blow to U.S. pharmaceutical companies that they were forced to announce "voluntary" cuts in prices for AIDS drugs to Africa by as much as 80%.[30] The statement of Donna E. Shalala, U.S. Secretary of Health and Human Services at the time, sent an unambiguous signal to PhRMA that the terms of the debate were shifting. Her statement noted that, "protecting intellectual property rights is fundamental to having a dynamic pharmaceutical industry, but at the same time we're recognizing the need to drag down the costs of drugs."[31] Though the executive order related only to Africa, its effects spilled over to the U.S./Brazil debate. Two days after the order, five major pharmaceutical companies agreed to negotiate price cuts in Brazil.

Another challenge of this USTR/PhRMA relationship is the multiple roles that the USTR has to play in balancing the conflicting interests of its principles. Too

---

[26] *Id.*, at 137.

[27] *Id.*

[28] An example of such a constituent may be a consumer group lobbying for cheaper drug prices in the U.S.

[29] May 10, 2000. Executive Order 13155, "Access to HIV/AIDS Pharmaceuticals and Medical Technologies."

[30] *New York Times, supra* note 8, at 18.

[31] *Id.*

much support for PhRMA upsets those constituents who believe that charging high drug prices for essential medicine in situations of health emergencies is unethical, since lower prices may save lives. Thus, on the one hand, the USTR's role is to pursue the interests of PhRMA, and on the other, to be responsive (indirectly, through the "elected" principal) to American NGOs demanding a trade policy that is much more accommodating of Brazil's interests. The agency costs to PhRMA can be translated into strategies to strengthen Brazil's negotiating power and to counter the benefits of this partnership.

8.4 Brazil's Strategy

According to common definition, "*negotiating power* is the ability to influence or move the decisions of the other side in a desired way."[32] Some scholars argue that what follows from this definition is that a more powerful partner negotiating with a weaker one will be able to achieve a more gainful outcome for itself. This implies that the USTR and the pharmaceutical companies—coming from the world's sole remaining "superpower" and with nearly unlimited resources at their disposal— would be able to achieve a more advantageous result than the Brazilian government. However, many negotiating cases indicate that the weaker party is often more powerful than at first perceived. To achieve a stronger position, weaker parties can pursue various strategies to increase their negotiating power.

Among a "lamb's" best strategies are "dependence," "autonomy" and "community," each of which are discussed in turn in this section.[33] Dependence is a bilateral strategy, autonomy is a unilateral strategy, and community is multilateral in nature. To achieve its goals of cutting prices for AIDS drugs and providing universally accessible treatment, Brazil pursued a combination of all three strategies, while capitalizing on agency issues embedded in the USTR/PhRMA relationship. Brazil attempted to align its interests with some principals that the USTR serves, used the U.S. domestic constituencies to lobby for change in the USTR's negotiating mandate and capitalized on the USTR's role conflicts.

8.4.1 Bilateral Strategy: Dependence

Dependence, in this context, is defined as "seeking support from other countries by creating an agreed upon dependency relationship."[34] The strategy of dependence is utilized mostly by smaller, less powerful states. But Brazil, a relatively powerful

---

[32] "How Should the Lamb Negotiate with the Lion? Power in International Negotiations" Jeswald W. Salacuse in *Negotiation Eclectics: Essays in Memory of Jeffrey Z. Rubin,* edited by Deborah Kolb, PON Books.

[33] *Id.,* at 91.

[34] *Id.*

developing country, used this strategy to gain the support of India, another influential developing country, in order to increase its power vis-à-vis the United States. The dependent relationship between Brazil and India was established bilaterally and was based on a realistic assessment of mutual gains. Through this relationship, India gained access to the Brazilian market for AIDS drugs ingredients that India manufactures generically. Brazil, in turn, was able to purchase cheap raw materials from which to produce end products—drugs used in the "triple cocktail." Most importantly, through this strategy, Brazil acquired an international ally in support of its cause.

8.4.2 Unilateral Strategy: Autonomy

At first glance, it seems that the strategies of dependence and autonomy are diametrically opposite. In Brazil's instance this is not the case, however, because the costs of the dependent strategy are not very high. Brazil's dependence on India is minimal, since it can unilaterally produce the raw materials it purchases from its Indian ally, only at a slightly higher cost. And even without Indian raw materials, the costs of manufacturing drugs locally in Brazil would still be significantly lower than the cost of purchasing these drugs from U.S. pharmaceutical companies. Thus the autonomy strategy strengthens Brazil's power relative to the United States, while making the costs of breaking up the relationship with India low.

One of the most effective ways Brazil autonomously increased its negotiating power was to develop a strong "best alternative to a negotiated agreement" (BATNA).[35] In developing its BATNA, Brazil used very effectively its available assets. These assets included international law and domestic intellectual capital, which were utilized to increase Brazil's self-sufficiency in producing AIDS drugs.

While Brazil's ability to acquire cheap components from India increases its bargaining power in relation to PhRMA, the alternative that increases Brazil's negotiating power even more is its ability to use "compulsory licensing" under the *ordrepublic* exceptions of the TRIPS agreement. Compulsory licensing is "an involuntary contract between a willing buyer and an unwilling seller imposed and enforced by the state."[36] Its use has a basis in general international law.[37] In the international legal context, compulsory licensing means the grant of a patent by the government for use without the permission of the patent holder, in situations

---

[35] See Roger Fisher, William Ury and Bruce Patton, *Getting to Yes: Negotiating Agreement Without Giving In*, 1991, PON.

[36] See Gianna Julian-Arnold, International Compulsory Licensing: The Rationales and the Reality, 33 IDEA 349 (1993), reprinted in International Intellectual Property Anthology (Anthony D'Amato and Doris Estelle Long eds., Anderson Publishing 1996) at 310 n.15.

[37] See John S. James, Compulsory Licensing for Bridging the Gap—Treatment Access in Developing Countries: Interview with James Love, Consumer Project on Technology. Available at http://www.immunet.org/imunet/atn.nsf/page/a-314-01.

when the patent holder is either not using the patent at all or not using it adequately.[38]

The TRIPS agreement attempts to strike a balance between the rights of private inventors and the public goals of government regulators. It requires all state parties to provide adequate intellectual property protections for patents, while allowing governments to compulsory license patented products under exceptional circumstances.[39] Both the United States and Brazil are signatories to this agreement.

Brazil has successfully argued before the U.S. that seizing patents may be justifiable in cases of extreme emergency, and that the national AIDS epidemic in Brazil is one such case. The USTR does not generally object to Brazil's right to issue compulsory licenses. The USTR, however, did oppose Brazil's interpretation of how this right should be applied. At the heart of the ongoing negotiations is the specific provision of Brazil's compulsory licensing law (Article 68, sec. (1)(I) that makes it legal for the Brazilian government to manufacture or import a generic version of a drug in cases when a patent holder fails to manufacture it in Brazil within the 3 years from the date when the patent was issued. Brazil has relied on domestic intellectual capital to develop local production capacity to generically manufacture patented drugs. Should the negotiations fail, Brazil still can issue compulsory licenses for AIDS drugs legally under international law and has the capacity to proceed with local manufacturing.[40]

Brazil's present BATNA has two major weaknesses, however. First, U.S. pharmaceutical companies are able to provide better distribution systems along with their products and to ensure the proper consumption of these drugs. Brazil could thus strengthen its BATNA by enhancing its distribution networks. Second, legal terms such as "working the patent" and "inadequate usage" (legal preconditions for the use of compulsory licenses)[41] are not clearly defined in the TRIPS agreement and are likely to produce further challenges to Brazil's domestic policies of fighting AIDS. This will lead to either further negotiations or legal challenges, which in turn may drain Brazil's resources and hinder its ability to be flexible in implementing its national health policies.

---

[38] Article 5, paragraph 2 of the Paris Convention states: "Each country of the Union shall have the right to take legislative measures providing for the grant of compulsory licenses to prevent the abuses which might result from the exercise of the exclusive rights conferred by the patent, for example, failure to work."

[39] See Duane, Nash, South Africa's Medicines and Related Substances Control Amendment Act of 1997, 15 Berkeley Technology Law Journal 486 (2000).

[40] See John S. James, Compulsory Licensing for Bridging the Gap: Treatment Access in Developing Countries: Interview with James Love, Consumer Project on Technology http://www.immunet.org/immunet/atn.nsf/page/a-314-01. Also see The Paris Convention for the Protection of Industrial Property of March 20, 1983, as revised, available at http://www.wipo.int/treaties/ip/paris/paris.html (accessed on November 11, 2001) and Agreement on Trade-Related Aspects of Intellectual Property Rights (TRIPS), Article 31, reprinted in 2000 International Trade Law Handbook 567, at 580.

[41] TRIPS, Article 31, *supra* note 14.

8.4.3 Multilateral Strategy: Using the International Community

To strengthen its power, a weaker "lamb" country may participate in international organizations that include the "lion," and make use of the strength of the community in order to achieve its goals.[42] The Brazilian government has pursued its case in many multilateral forums, such as the United Nations (UN), the World Health Organization (WHO) and the WTO. Some of Brazil's tactics have included international consensus building on important issues such as differential pricing; alliance and coalition building; and "attention getting" in appropriate forums. Each of these tactics are discussed below.

Brazil has been successful in building international support for differential pricing, which is another alternative to purchasing drugs at high prices. Differential pricing is a two-tiered structure whereby countries with a high level of economic development pay a higher price for drugs and countries with lower level of economic development pay a lower price for the identical drugs.[43] Brazil has participated actively in workshops on differential pricing organized by the WTO and WHO secretariats.[44] These workshops have brought together experts from various fields to think creatively about solutions, and have helped to create an emerging international consensus on the issue.

The participants at one WTO/WHO workshop seemed to agree that differential pricing could, and should, play an important role in ensuring access to existing essential drugs at affordable prices in developing countries, while allowing the patent system to function effectively so as to provide incentives for the development of new medicines. One of these workshops also set up "a framework for dialogue," which suggested that sporadic, case-by-case negotiations with pharmaceutical companies for price cuts do not present a viable solution to the problem. The workshop concluded that differential pricing enabled countries to develop "a more systematic approach…than is possible through ad hoc discounts offered at the discretion of individual companies."[45] Brazil's participation in these workshops was very important. The Brazilian government used them as forums for strategic advocacy, in which the country was able to convince the U.S. to accept a

---

[42] Salacuse, *supra* note 32, at 92.

[43] See generally "Workshop on Differential Pricing and Financing of Essential Drugs," Background Note Prepared by Jayashree Watal, Consultant to the WTO Secretariat; Report of the Workshop on Differential Pricing and Financing of Essential Drugs, World Health Organization and World Trade Organization Secretariats, Norwegian Foreign Affairs Ministry, Global Health Council, 8–11 April 2001, HØsbØr, Norway. Available at http://www.who.int/medicines/docs/par/equitable_pricing.doc and http://www.wto.org/english/tratop_e/trips_e/wto_background_e.doc.

[44] *Id.*

[45] *Id.* For example, Jean-Pierre Garnier, GlaxoSmithKline's chief executive, made a statement of intent to the effect that the company would place an increased emphasis on differential pricing, Higel Cope, "GlaxoSmithKline Proposes a Two-Tier Drug Pricing Structure," http://www.independent.co.uk/story.jsp?dir=1&story=68564&host=1&printable=1 (Accessed on November 1, 2001).

more flexible negotiation formula. As a result, the two parties agreed on a formula that holds that IPR protection should not be pursued at the expense of human lives.

Past cases have demonstrated that it is possible to persuade a more powerful state to make concessions by forming coalitions against that state.[46] One effective tactic that Brazil used was to build cooperative relationships with third parties sympathetic to its cause. In this way, Brazil clearly exploited the "multiple principals" problem faced by the USTR. Brazil involved AIDS and human rights NGOs, as well as African countries fighting the AIDS epidemic. This was an effective "negative strategy"[47]—it reduced reward for the United States by increasing the costs of aggressively pursuing the issue.

Brazil was very successful in attracting the attention of activist groups and other NGOs in both Brazil and the United States. These organizations took up Brazil's cause and worked hard toward its advancement, particularly in the area of dissemination of information. For example, it was fairly common to come across articles (in newspapers and on the internet) with headlines such as "Unethical Patent Law: How the United States and the WTO Impact the Health of Brazilian Citizens."[48] These types of articles, and the frequency with which they appeared in the media, helped to sway world public opinion towards the Brazilian side. As a result, the U.S. position began to be perceived as siding with the pharmaceutical companies in their blind conquest for profits and hindering Brazil's struggle to save human lives.

Among the most outspoken critics of the U.S. policy was the Nobel Prize-winning group Medicins sans Frontieres (Doctors without Borders), which warned that the U.S. challenge in the Dispute Settlement Body (DSB) of the WTO "might handicap the successful Brazilian AIDS program, which [was] largely based on Brazil's ability to manufacture affordable treatment."[49] While it is difficult to assess the amount of influence these groups had on the formulation of U.S. policy, NGO criticism is said to have contributed substantially to the U.S. decision to withdraw the case from the WTO's DSB.

In addition to allying with these groups, Brazil has been very supportive of African countries asserting their right to affordable medicine. "Brazil has raised this banner because it is a cause that has to do with the very survival of some countries, especially the poor ones of Africa," said President Cardoso of Brazil in an interview with the *New York Times*. "This is a political and moral issue, a truly

---

[46] Jeffrey Z. Rubin and Jeswald W. Salacuse, "The Problem of Power in International Negotiations," in International Relations, at 32. Successful coalition efforts include OPEC and ASEAN.

[47] *Id.* at 28.

[48] "Unethical Patent Law: How the United States and the WTO Impact the Health of Brazilian Citizens," Article #46, published by Free Information Property Exchange; http://www.freeipx.org/display.php3?id=46 (Accessed November 9, 2001).

[49] Bureau of National Affairs, "Pharmaceuticals: United States Drops WTO Case Against Brazil Over HIV/AIDS Patent Law," at 1. Accessed on November 20, 2001: http://www.cptech.org/ip/health/c/brazil/bna06262001.html.

dramatic situation, that has to be viewed realistically and can't be solved just by the market."[50] Brazil even offered to transfer its technology to these countries in order to facilitate the manufacture of generic AIDS drugs and save lives. In addition to making new allies in its confrontation with the United States, this strategy has demonstrated Brazil's goodwill and forced public opinion to see Brazil's strategy in a positive light. Because of the negative publicity directed at the United States from Brazil's allies around the world, a U.S. victory, if achieved, could be denounced as an irresponsible use of power.

By allying with international organizations, NGOs and other civil society actors, Brazil has been able to attract world attention to the issues of special importance to it. Perhaps the biggest "stage" where it received such attention— and one of the most strategic and important forums in these negotiations—was the UN Special General Assembly (GA) session in June of 2001.

GA sessions have traditionally served as forums in which developing countries can push forward their agendas. During the special GA session dedicated to AIDS issues, Kofi Annan, then Secretary-General of the UN, stated that, "the sheer magnitude of the 'human tragedy' had compelled global action."[51] The resolution that came out of the special session commits governments to "cooperate constructively in strengthening pharmaceutical policies and practices, including those applicable to generic drugs and intellectual property regimes."[52] While this statement can be interpreted as both affirming the preservation of existing patent policies and allowing more access to essential medicines for Third World poor, it put a great deal of pressure on the U.S. and contributed to that country's decision to withdraw its case from the WTO. Thus, even though GA sessions rarely lead to tangible results, Brazil was able to capitalize on the worldwide attention the conference brought to the AIDS subject, as well as Brazil's successful national policy of fighting the pandemic. The session made it difficult for the U.S. to win the public relations battle at home or abroad.

8.5 Lessons Learned and Future Strategies

Brazil is now undertaking bilateral negotiations with the U.S. armed with more than a decade of negotiation experience in this dispute. While during that time Brazil has generally been able to maintain its ground and has been successful in its attempts to advance the cause, some opportunities have been missed, and lessons can be learned and used in the ongoing negotiations. Some of the key lessons that

---

[50] *The New York Times*, March 31, 2001 "Maker Agrees to Cut Prices of 2 AIDS Drugs in Brazil, available at http://www.nytimes.com/2001/03/31/health/31AIDS.html?pagewanted=print.

[51] *The New York Times*, "The UN Members to Report Progress Toward Reducing Spread of AIDS," A10, Thursday, June 28, 2001.

[52] *Id.*

can be learned from Brazil's negotiations with the USTR and PhRMA are as follows.

1. **Political Will**: A key determinant of Brazil's effective response to the AIDS epidemic has been its government's strong commitment. The Brazilian government enacted policies that increased resources for heath care, created a national AIDS prevention and treatment program and reduced the price of AIDS drugs. This strong commitment helped strengthen the Brazilian negotiators and assisted them in formulating creative alternatives and options. The political will of the Brazilian government also helped to attract key partners from civil society to support its cause and to mobilize world opinion in its favor.

2. **Framing the Issue**: Brazil framed the issue as one of "access to essential medicines," instead of the narrower "reducing the price of AIDS drugs." In so doing it was able to attract diverse partners to support its strategy and build a strong coalition. This also helped Brazil to weaken the USTR's negotiating position, since it revealed that the USTR was only pursuing PhRMA's narrow interests in the negotiations.

3. **Exploiting Agency Costs**: Since PhRMA was negotiating through the powerful USTR, Brazil successfully exploited the conflicts inherent in this kind of "principal-agent" relationship. Brazil influenced the other principals of the USTR, whose interests were not aligned with PhRMA's, in order to exert pressure on the U.S. government. It also weakened the position of PhRMA by helping to change the mandate of the USTR. Furthermore, by framing the issue broadly, Brazil helped to perpetrate role conflicts within the USTR.

4. **A Strong BATNA**: Brazil strengthened its BATNA through a multi-level approach. The country established a bilateral strategy with India to import raw materials for the drugs, making its threat to produce drugs locally more credible. It established a multilateral strategy in which it built strong coalitions (with international funding agencies, civil society actors, international organizations and the community of nations) to support its cause of having access to medicines to save lives. It also played a very active role in developing alternative strategies, like differential pricing, that weakened PhRMA's position. This also helped it gain widespread support in the court of public opinion. Lastly, Brazil pursued a number of unilateral policies designed to exploit ambiguities in the TRIPS agreement dealing with compulsory licensing.

Before the U.S. withdrew its case from the WTO, Brazil was able to counter the United States' enormous bargaining power through a combination of tactics and strategies. Brazil attempted, and to a certain degree succeeded, in building international consensus on issues that were important for Brazilian citizens by participating in many forums and workshops such as those dealing with differential pricing. Brazil was also successful in its efforts to create coalitions with various groups and countries with which it had common interests, such as human rights groups and African countries also affected by AIDS pandemic. Furthermore, Brazil made good use of prestigious international forums such the UN General Assembly to draw attention to its cause. Through a combination of such strategies and tactics, Brazil was able to augment its negotiating power and hold its ground against a more powerful party.

By following Brazil's lead and carefully building coalitions and partnerships, persistently disseminating information, capitalizing on public support and intelligently choosing alternatives to a negotiated solution, a future weaker "lamb" will surely have a better chance to hold its ground against a "lion."

8.6 Addendum

Though bilateral negotiations between the U.S. and Brazil continue, Brazil's AIDS program today produces generic drugs locally, thus decreasing the cost of AIDS medication and allowing the government to provide AIDS medications to those most in need. Moreover, its relationship with India, its autonomy from the United State and its support for poor African nations seeking affordable treatment of their own have all won Brazil many allies and helped it reshaped the global response to HIV/AIDS.

Brazil's leadership has led to an international challenge to intellectual property rules, as part of an effort to make AIDS treatment affordable globally. A Brazil-sponsored WHO resolution in 2003 on "Intellectual Property Rights, Innovation, and Public Health" led to a 2005 report of the same title, which finds that IPRs have not been helpful in addressing public health issues (Nunn 2009, p. 143). Then in May 2006, Kenya joined Brazil in sponsoring another resolution that created the Intergovernmental Working Group on Innovation, Intellectual Property, and Public Health (IGWG), which led to a 2008 resolution entitled "Global Strategy on Public Health, Innovations and Intellectual Property." These resolutions and ongoing work by the WHO, which "will appoint a working group to make policy recommendations for implementation of the strategy by 2010" (Nunn 2009, p. 143), demonstrate a global affirmation of Brazil's decision to produce drugs generically and locally rather than buying patented drugs from pharmaceutical companies in the United States.

Brazil's National AIDS Program has successfully met the challenge of providing affordable AIDS treatment, and its program has kept infection rates low. At present 660,000 people in Brazil are infected, a sizeable number to be sure, but only 0.7% of the population and "only half the number predicted by the World Bank a decade ago" (Nunn 2009, p. 11; Reel 2006). Former President Fernando Henrique Cardoso reports that "Today, 185,000 people receive life-saving AIDS cocktails in Brazil, and thousands of lives have been saved" (Nunn 2009, p. ix). Brazil may be losing its competitive edge in the generic drug industry, but its relationship with India has helped it weather the changing market. Now it imports certain drugs from generic drug companies in India (Nunn 2009, p. 156). Moreover, "when Brazil's broader AIDS treatment institutions are considered, accounting for the relatively more costly locally produced generics and Brazil's reduced costs from price negotiations, Brazil still saved nearly $1 billion from 2001 to 2005" (Nunn 2009, p. 156). In light of these observations, Brazil's AIDS program seems remarkably effective.

For the most part, Brazil's health negotiations seem to be a success; it met its main interest—making AIDS medicine affordable and accessible for its citizens—and it reinforced good relationships with many countries globally and has established itself as an important player in the global public health debate.

However, Brazil's relationship with the United States remains a challenge. Not all of this stems from the specific issue of generic versus patented drugs; ideological differences have also made subsequent negotiations difficult. In 2003, an amendment to U.S. law H.R. 1298 (authorizing the President's Emergency Plan for AIDS Relief) stipulated that U.S. funds can only be donated to groups "explicitly opposing prostitution and sex trafficking" (Ribando 2007, p. 22). In addition, the U.S. insists on an abstinence-only AIDS prevention policy. These policies clash with those of Brazil, which respects its sex workers and wants them and their clients to be protected. In fact, sex workers are big advocates of Brazil's AIDS prevention program, and they carry government-issued condoms with them (Reel 2006). One of the sex workers, Paula Duran, was quoted as saying, "I'm always telling people that they should never do anything without a condom. A lot of the young people who come around here don't know anything about it, so I try to teach them whatever I can" (Reel 2006). Brazil does not want to change its successful program because of new U.S. restrictions. Sonia Correa, an AIDS activist and co-chair of the International Working Group on Sexuality and Social Policy, accused the U.S. of "bullying, pushing and forcing" in 2005 when the U.S. offered $40 million to Brazil for an abstinence program that did not treat prostitutes. Brazil turned down the money (BBC 2006). It is unfortunate that Brazil did not receive the aid, but they seem to be doing fine without it. The United States, on the other hand, has further isolated itself from Brazil and reduced its influence in the region through the pursuit of such policies.

This ideological dispute is arguably a separate issue from the 2001 negotiations, which focused on intellectual property and general access to AIDS treatment rather than the specific nature of its use. Still, any bitterness the U.S. felt from those negotiations certainly did not increase its likelihood to work through ideological differences. That is not to say that there is no hope. Clare Ribando's report, written in 2007, describes relations between the two countries as "fairly warm and friendly" (Ribando 2007, p. 1). And there are signs that relations can improve further with Barack Obama now as President, considering Obama's developing friendship with Brazilian President Luiz Inacio Lula da Silva, (known as "Lula"). Indeed President Lula says that Obama's presidency offers "an opportunity for Latin America to build a relationship with the United States that it did not have before" (Earth Times). Whether this improved relationship between the presidents of the two countries can help Brazil and the United States work through their disagreements over issues as diverse as intellectual property and prostitution remains to be seen.

Though Brazil will continue to face challenges, the successful results of its 2001 negotiations illustrate the importance of basing a negotiation strategy on an understanding of each party's interests. And indeed, it can be said that Brazil's strategy has created value for the entire international community. While the United States, under the Bush Administration, missed an opportunity to play a role in the emerging generic drug market, the improved relationship between the two countries suggests room for future growth. In the meantime, Brazil must continue to respond to changes in the global generic pharmaceuticals market and to the challenges that come from success, namely that AIDS patients are living longer, increasing the price of their care. Brazil seems capable of meeting the new challenges, if its successful AIDS program is any indication of future success.

Chapter 9
Case III—Keeping Your Head Above Water in Climate Change Negotiations: Lessons from Island Nations

Abstract International negotiations to address the threat of global climate change began nearly 15 years ago and have proceeded almost continuously since then. These climate change negotiations have been among the most complex multilateral negotiations ever conducted internationally. Stuart Eizenstat, a former US Under Secretary of Commerce and the chief US negotiator at the Kyoto Protocol negotiations, once commented, "Few issues are as cross-cutting and politically charged as climate change, involving energy use, land use, [and] a wide variety of industrial and agricultural concerns" (Lakshmanan 1997).

Keywords Global climate change negotiations · Kyoto Protocol · AOSIS · Intergovernmental Panel on Climate Change (IPCC) · NGOs · Group of 77 · Group of 77 and China · Coalition building · Umbrella group · Intergovernmental Negotiating Committee (INC) · Framework Convention on Climate Change · Conference of parties (COP) · Greenhouse gases · Negotiation process · Good chairman qualities · Issue trade-off · Preparation for negotiation · Single text process · Institutional capacity-building · Power · Issue linkage

This case focuses on the strategies and tactics of negotiators from the Alliance of Small Island States (AOSIS) during the climate change negotiations. The AOSIS representatives were highly effective in making their countries' concerns heard and in meeting their national interests. Their experience shows how negotiators from countries with relatively little political or economic power can use moral persuasion and skillful negotiation tactics to influence more powerful countries in complex, multi-issue, multi-party international negotiations.

Case study written and researched by Kelly Sims Gallagher and Allison Berland.

D. Fairman et al., *Negotiating Public Health in a Globalized World*,
SpringerBriefs in Public Health, DOI: 10.1007/978-94-007-2780-9_9,
© The Author(s) 2012

9.1 The Threat and Causes of Global Climate Change

Swedish physicist Svante Arrhenius postulated the theory of the greenhouse effect in the late 1800s. But his theory was not substantiated until the 1960s, when researchers in Mauna Loa, Hawaii, documented a steady increase in the amount of carbon dioxide (CO_2) in the atmosphere. CO_2 is a "greenhouse gas" because it traps heat from the sun in the earth's atmosphere. At natural levels, heat-trapping gases such as CO_2 maintain the earth's habitable climate, but when humans release too many greenhouse gases into the air the climate begins to change. During the Industrial Revolution, countries now considered "industrialized" burned enormous quantities of coal and other fossil fuels, thereby releasing excessive amounts of CO_2 into the atmosphere. Greenhouse gases like CO_2 stay in the atmosphere for about a century, trapping heat that would otherwise be released. As a result, average global temperatures have already increased, with substantial warming in some parts of the world.

The consequences of this greenhouse effect are profound and will differ around the globe, depending on regional conditions. Anything that depends on the earth's climate system will be affected, from weather patterns to crop production to human health. Scientists from the Intergovernmental Panel on Climate Change (IPCC), the official body of scientists advising world governments on the state of climate science, have warned of more extreme weather in the form of heat waves, floods and droughts. Water supplies and agricultural production may be curtailed in some parts of the world, and human health could suffer from a wider spread of infectious diseases. Forests and other ecosystems will be forced to adapt quite rapidly to climatic conditions, or perish. New extinctions of plant and animal species are likely, especially in areas most severely affected. Ocean temperatures will rise, causing coral reef bleaching and sea-level rise. Sea-level rise is already occurring, primarily because of thermal expansion of ocean water (due to higher ocean temperatures) and also from melting mountain glaciers. Developing countries are likely to be more adversely affected by these changes than industrialized countries, because the poor will be least able to adapt, given the high costs associated with coping with these threats. There are still uncertainties about exactly when climate change will occur and how severe the impacts will be, but there is broad scientific consensus that the greenhouse effect exists, that humans have altered the natural climatic balance and that continued greenhouse gas emissions will be increasingly destabilizing for the climate (IPCC 2001a).

The main heat-trapping gases are CO_2 (which accounts for the majority of the world's emissions), methane (CH_4), nitrous oxide (N_2O), hydrofluorocarbons (HFCs), sulfur hexafluoride (SF_6) and perfluorocarbons (PFCs). The first two gases mostly come from the burning of fossil fuels, such as coal in power plants or gasoline in cars. Nitrous oxide is released from agricultural soils, cattle feed lots and the chemical industry. Hydrofluorocarbons are used mostly for refrigeration, and perfluorcarbons are used in industrial applications. One perfluorocarbon, perfluoromethane (CF_4), resides in the atmosphere for at least 50,000 years (IPCC

2001b). Sulfur hexafluoride is a fairly rare chemical that also has an extremely long life in the atmosphere. All of these gases will be regulated by the Kyoto Protocol, which is the most recent international treaty on this subject—if it enters into force.

Industrialized countries have emitted the vast majority of the world's emissions during the last century. One study estimates that, since 1950, industrialized countries have emitted about 85% of the CO_2 already in the atmosphere (Sari 1998). Today, the United States accounts for one quarter of the global emissions of greenhouse gases. On a per capita basis, industrialized countries are much more greenhouse-gas-intensive, with the average American citizen emitting 20 times more than an Indian citizen, for example, and 10 times more than a Chinese citizen. Because it is a rapidly developing and populous country, China is now the second-largest overall emitter of greenhouse gases. Other rapidly industrializing countries are increasingly becoming substantial contributors to the greenhouse effect as well.

Once scientific concern about climate change was well established, and environmental nongovernmental organizations (NGOs) began to express a desire to take action to address the threat, a multilateral negotiation process was initiated.

9.2 Climate Change Negotiations, Pre-Kyoto

In response to scientific research and increasing public concern about possible human disruption of the global climate system, the United Nations (UN) adopted Resolution 43/53 in 1988 on the "protection of global climate for present and future generations of mankind." That same year, the first international meeting was held that brought scientists and governments together to discuss taking action on climate change. This meeting was called the "Toronto Conference on the Changing Atmosphere." At this conference, industrialized countries' governments pledged to voluntarily cut their CO_2 emissions by 20% by the year 2005. This later became known as the "Toronto Target," and it was formally adopted and proposed by the AOSIS countries in 1994.

Also in 1988, the World Meteorological Organization (WMO) and the United Nations Environment Program (UNEP) established the Intergovernmental Panel on Climate Change to review existing scientific research on climate change and provide recommendations on future action. In 1990, the IPCC published its first "assessment report," which concluded that climate change was a real threat that should be taken seriously, further motivating the countries to formulate an international treaty to address the issue.

Because so many countries were involved in negotiations over the treaty, coalitions of like-minded countries formed. It turned out that many groups of countries shared similar interests and wanted to negotiate together to aggregate their power. As a practical matter, these negotiation blocs made the actual negotiations somewhat simpler because, in effect, the number of parties was reduced;

one coalition representative could negotiate on behalf of a much larger group of countries.

Although the coalitions have changed a little over time, the main ones that emerged at the beginning have endured. All of the developing countries are part of the "Group of 77 and China" (G-77 and China). Among these developing countries, more specific coalitions formed as well. The oil-producing states (OPEC) became a negotiating bloc that was generally resistant to the creation of a legally binding treaty to address climate change. In 1990, the Alliance of Small Island States was formed out of concern for the survival of their countries, because of the threat of sea-level rise. AOSIS now works on many other issues besides climate change, but it was initially established to become a negotiating force in the climate change negotiations. Several alliances of developing countries in particular regions also formed.

The industrialized countries organized into three main coalitions. Members of the European Union (EU) formed one, and countries with economies in transition (EIT) created another. The rest mostly converged into a group eventually known as the "Umbrella Group." The Umbrella Group had about a dozen members, including Japan, the United States, Canada, Australia and New Zealand.

The AOSIS countries joined together because they recognized that they all faced a unique threat in the form of sea-level rise. Low-lying island states by their very nature can easily be swamped and rendered completely uninhabitable by only a few inches of sea-level rise. Most of the low-lying island countries were part of the British Commonwealth, but the president of the Maldives, Maumoon Abul Gayoom, realized that there were probably too many different points of view in that association of countries and that a more focused group of countries with exactly the same interests was needed. He observed in 1999, "We are a very low-lying country and could face serious problems due to rising sea levels in the coming generations. There is very little a small country can do to overcome such a serious global problem. So we have to have international understanding and cooperation in order that we and other countries like us in the world can cope with such problems as sea levels rising, climate change and global warming."

In 1989, expert scientists were brought to a conference called the "Small States Conference on Sea Level Rising," which was convened in the Maldives to brief representatives from other island countries about the threat. At that conference, President Gayoom succeeded in persuading other island countries to join AOSIS. It is interesting to note that the Maldives does not produce or consume significant amounts of commercial energy; it imports only 2,000 barrels of oil per day for all of its energy supply (EIA 2001). Eighty percent of its territory sits less than three feet above sea level, which is within IPCC projections of sea-level rise for the next century.

In December 1990 one AOSIS member, Malta, sponsored Resolution 45/212 in the UN General Assembly, calling for the establishment of an Intergovernmental Negotiating Committee (INC) that would have a mandate of creating a new Framework Convention for Climate Change. During the next year and a half, the INC sought to find consensus among the 150 participating states (both industrialized and developing nations) in drafting a new treaty framework to address the threat of climate change.

Among the issues that arose immediately were who was responsible for the problem, what kind of treaty should be negotiated, whether or not it should be legally binding and exactly how countries would reduce their emissions. Developing countries raised many equity concerns; they pointed out that wealthy countries had industrialized using energy-intensive manufacturing industries, which had emitted most of the CO_2 and other greenhouse gases already in the atmosphere. Developing countries felt that they also had a right to develop any way they determined necessary, and that industrialized countries should take initial responsibility for addressing the problem and reducing emissions (although developing countries conceded that they should do what they could as well). Industrialized countries have contended since then that they did not know about the climate change problem during their industrialization, and that some developing countries are major emitters of greenhouse gases today; so, to truly address the problem, all countries should take concerted action together to reduce heat-trapping gases.

AOSIS was formally convened as a negotiating group in time for the first meeting of the Intergovernmental Negotiating Committee in February 1991. During the INC negotiations, AOSIS began to adapt to the more formal UN context of interaction. They began to meet as a coalition on a regular basis during the actual negotiation sessions. They formed a list of core principles that guided their negotiation, such as the principle of precautionary action, the precautionary principle, duty to cooperate, the polluter pays principle and state responsibility, equity, common but differentiated responsibility, and energy conservation and development of renewable energy sources (SIDS 2001). They solicited experts from NGOs and other scientists to brief and assist them in formulating legal language and in understanding the science and policy options for addressing the problem. For example one NGO, the Foundation for International Environmental Law and Development (FIELD), provided legal advice to AOSIS that enhanced its ability to propose legal language and actual negotiating text. Over time, AOSIS formed relationships with some experts that have endured to this day.

Negotiations on the INC draft—at this point called the UN Framework Convention on Climate Change, or UNFCCC—culminated in a high-level negotiating session during the United Nations Conference on Environment and Development (UNCED) at Rio de Janeiro in 1992. The UNFCCC had been negotiated from 1989 to 1992, and became international law in 1994 when 50 countries ratified it. The UNFCCC's main objective is "stabilizing atmospheric concentrations of greenhouse gases at safe levels." A number of important principles were established in the UNFCCC, and a reporting and inventory system was set up to monitor the world's emissions of greenhouse gases (GHGs). No binding emission-reduction commitments are contained in the UNFCCC, but industrialized countries formally accepted responsibility to take the lead in reducing emissions. They promised to try to reduce their greenhouse gas emissions to 1990 levels by the year 2000.

Soon after the UNFCCC was finalized, AOSIS realized that since the emission-reduction targets in the framework convention were voluntary, a treaty with mandatory reductions was likely to be needed. They began to analyze how much greenhouse gases would have to be reduced to completely avert the threat of climate change, how much these gases could be reduced without harming the world economy, and what kinds of reductions might actually be politically feasible. They formulated a collective position based on their answers to these questions and formally proposed a target for emissions reductions in September 1994: They advocated a 20% reduction below 1990 levels by 2005 for industrialized countries. "It was a deliberate position to get this out on the table. There was nothing at all in the convention except a very generalized obligation...," said Ambassador Tuiloma Neroni Slade, Permanent Representative of Samoa to the United Nations and Chairman of AOSIS. In putting forth a proposal with an ambitious target and timetable, AOSIS hoped to raise expectations of what a future protocol could include.

This position was mostly rooted in science. The IPCC had determined that in order to prevent climate change, emissions would have to be immediately reduced by close to 80%. At the earlier meeting in Toronto in 1988, however, a 20% reduction had been viewed as reasonable from a political point of view, and AOSIS had also seen studies that estimated that such reductions were economically feasible. Moreover, they felt that the target was clear, easy to understand and implementable, and it had environmental integrity. The environmental NGO community soon endorsed this proposal and held to this endorsement throughout the next 4 years of negotiations. AOSIS's submission was the first legally binding proposal to be formally introduced in a UN setting; it therefore provided the initial framework for all subsequent negotiations.

After the UNFCCC entered into force, the first meeting of the Conference of Parties (COP) was held in Berlin in 1995. This meeting coincided with the release of the second assessment report of the IPCC, which declared that there was a "discernible human imprint on the global climate." In other words, scientists were reasonably convinced that humans were disrupting the climate. This report created a huge amount of press coverage and provoked fresh concern among NGOs and many previously unconvinced politicians around the world.

By the time of the Berlin meeting, scientific evidence, public opinion and political activism had increased pressure on industrialized countries to make binding greenhouse gas emissions reductions. In response, the countries negotiated a mandate for themselves, the Berlin Mandate, which set a deadline of December 1997 for the negotiation of a legally binding protocol to the UNFCCC that would establish specific targets and timetables for emissions reductions. This Mandate reiterated industrialized countries' responsibility for acting first to reduce emissions. All countries present in Berlin, including the United States, adopted the Berlin Mandate. The eventual outcome of the ensuing negotiations was the adoption of the Kyoto Protocol in December 1997.

After the Berlin meeting, the negotiations for the Kyoto Protocol were mostly conducted by an open-ended Ad Hoc Group on the Berlin Mandate, known as the

AGBM, which met frequently and was a less-formal body than the Conference of Parties. However, the COP also continued to meet once a year. The AGBM met eight times between August 1995 and December 1997. During these 2 years, countries debated the form and content of the future Kyoto agreement. In addition to the primary question of what the result of the negotiation would be—a protocol, an amendment or some other legal instrument—other key issues related to the size of emission-reduction targets, the timetables, which greenhouse gases to include in the agreement, and whether or not the policies and measures countries would employ to reduce emissions would be recommended or mandatory.

AOSIS viewed the AGBM negotiations as essential, because finally the world community was supposed to create a mechanism for obligatory emissions reductions. AOSIS wanted to push for as strong a target as possible and was also determined to make sure the protocol would be legally binding and have environmental integrity. The political conditions seemed favorable, with the Clinton-Gore Administration's understanding of the climate change threat in America and environmental political movements rapidly gaining momentum in Europe. On the other hand, very serious challenges remained, including the potential blocking power of the OPEC states, powerful political forces arrayed against action from fossil fuel producers in the United States and other industrialized countries, potential divisions among the G-77 and the risk of deadlock between industrialized and developing countries about the level of reductions and the distribution of responsibilities.

9.3 The Kyoto Protocol Negotiations

Before the AGBM negotiations even began, AOSIS prepared for them by engaging in internal consultations and developing briefing papers. Using a comprehensive briefing document based on their 1994 proposal, AOSIS entered the negotiations prepared to defend their public positions, while also understanding they would have to be flexible on the floor. AOSIS believed it was important to have a common briefing document to serve as a guide, because it would keep people "on the same page," but this was balanced with the obvious need to be open to changes in real time.

Prior to the Kyoto talks, many of the AOSIS diplomats were stationed in New York at their countries' missions to the United Nations, so they were able to consult and meet in person. In addition, they had cultivated scientific and policy expertise over time and had begun to resemble a group of seasoned diplomats. By 1995, several AOSIS representatives had been working full time on the climate change issue for years. Some of them had prepared the in-depth national communications report about their country's emissions to the Secretariat of the UN-FCCC, for example, and learned a lot about the structure of AOSIS country emissions in the process. Others had worked for the UN Secretariat, the

Organization for American States, and The Caribbean Community (CARICOM), learning more about how to work within multilateral settings.

During the AGBM negotiations, AOSIS made a point of being at least as prepared as any other developing country delegation. As Marshall Islands representative Espen Ronneberg noted, this preparation gave AOSIS "a lot of influence politically," because they were fully prepared to get into the substance of the negotiations. Their counterparts from other countries easily ignored diplomats who came unprepared to the negotiation sessions. Thorough preparation thus helped to enhance AOSIS members' credibility in the negotiation process. Mr. Ronneberg's own credibility as a leading negotiator helped him to be elected Vice President for both the Third and Fourth Conferences of Parties to the UNFCCC, as well as Co-Chairman of the Joint Working Group on Compliance with the Kyoto Protocol. The Third Conference of Parties was the actual negotiation body for the Protocol in Kyoto, Japan.

Even though diplomats from the 186 countries participating in the negotiations had worked hard to narrow disagreements during the AGBM negotiations leading up to the Kyoto summit, many of the contentious issues remained unresolved right up until the final negotiations in Kyoto, Japan, at the end of 1997. The main disagreements had to do with the level of the targets, the timetables, the mechanisms for emissions reductions (such as emissions trading and joint implementation), whether or not carbon "sinks" such as forests could be counted against emissions, and whether or not developing countries would have to make binding commitments to limit their emissions of greenhouse gases.

During the sixth and seventh sessions of the AGBM, proposals from the major negotiating blocs were formally submitted, and a draft framework text was created from the areas of agreement. By the end of the AGBM negotiations, agreement had been reached on a Preamble, the Secretariat and some institutional elements of the Protocol (Oberthur and Ott 1999). AGBM Chairman Raul Estrada-Oyuela of Argentina created a draft protocol to be used as the text under negotiation for the start of the Third Conference of the Parties.

Thousands of participants came together in Kyoto in December 1997 to take part in negotiations that led to the adoption of the Kyoto Protocol. The Kyoto summit included 2,200 delegates from almost 200 parties to the UNFCCC, 4,000 observers from NGOs and 3,700 media representatives (Oberthur and Ott 1999). The 11-day summit was chaired by Estrada-Oyuela, whose strong leadership (at times forcefully reminding participants of their mandate), credibility as a developing country representative, trust from negotiators and intimate knowledge of the positions of the key players kept the process moving forward.

The AOSIS Chair, Ambassador Tuiloma Neroni Slade, had carefully assembled a relatively large and experienced group of negotiators to represent the AOSIS coalition. Having a large group was essential, because Ambassador Slade anticipated that there would be many concurrent formal and informal negotiating sessions and thought it was essential to have representatives at all of them. Internal coordination was a constant challenge, but Ambassador Slade exhibited "good Chairman qualities" by being able to "run good meetings, be an empathetic listener, understand the issues, keep things moving and manage internal disagreements," according to one observer. Meanwhile Marshall Islands representative Espen Ronneberg from AOSIS had secured the position of Vice Chairman of the Conference of Parties during these negotiations. This provided him with a means to guide the discussions, because he represented AOSIS on the Bureau of the COP. The Bureau was important because the Chairman would often sound out ideas on the Bureau before formally proposing them to the entire group of 186 countries.

During the Kyoto summit, AOSIS met every morning to review developments from the day before, plan for the upcoming day and distribute responsibilities. They developed a team approach for each major issue under debate. This meant that a team of people would be responsible for handling the same issue (such as emissions trading or emissions targets) throughout the Kyoto negotiations. This enabled AOSIS to match larger delegations from the EU, the United States and OPEC, all of which had experienced diplomats who knew particular issues in great detail. Each AOSIS team acted as the coordinator of the particular issue for the entire AOSIS coalition, and they were responsible for negotiating faithfully on behalf of their colleagues. In order to enhance trust in each other, Ambassador Slade tried to make sure that there was regional representation on each team. As one member observed, "having a flexible and friendly group is helpful."

During the Kyoto talks, Ambassador Slade not only had to manage the AOSIS coalition itself and represent AOSIS at all high-level sessions, but he had to play a leadership role within the G-77. According to one observer, Ambassador Slade had "no qualms about asserting himself" when necessary to gain access to important meetings or to convince others to hold to an agreed-upon position. At G-77 meetings, AOSIS would enunciate its point of view and look for countries that were attracted to their viewpoint. When necessary, Chairman Slade would remind his colleagues that entire countries were at risk of being eradicated, if aggressive action were not taken. AOSIS believed that, when possible, it was more advantageous to work as a block of 140 (i.e., the entire G-77) than as 43. "You had to speak up and defend your point of view, but at times we modified our position," said Ambassador Slade.

Keeping the G-77 constructively in alignment with AOSIS was a major challenge for Ambassador Slade. The G-77 tended to be weak about whether or not the Protocol needed to be legally binding. Some countries were doubtful about the need for action, and others were most concerned about the potential costs to them of having to reduce emissions. Although some AOSIS countries were willing to take on emission-reduction commitments, AOSIS decided to stick with the G-77 position that developing countries should not have to take on legally binding reductions until after the industrialized countries had proved themselves. Although it wasn't a deliberate trade-off, AOSIS managed to hold the G-77 to the concept of mandatory emission-reduction responsibilities for the industrialized countries.

The final Kyoto Protocol to the UNFCCC was adopted on December 11, 1997. It requires industrialized countries to reduce their emissions by 5% below 1990 levels by 2010, along with establishing new international mechanisms for meeting these targets. These mechanisms include international emissions trading, joint implementation and the clean development mechanism. These three mechanisms are intended to reduce the economic costs of emission reductions. All six main greenhouse gases are regulated under the protocol, and countries are allowed to offset their emissions with carbon sinks such as forests to a limited degree.

The final Protocol meets many of AOSIS's primary interests. Most importantly, the Kyoto Protocol is legally binding and not voluntary. AOSIS believed that voluntary commitments could not be counted on, especially since almost every industrialized country was about to fail to meet its voluntary UNFCCC target of 1990 levels by 2000. Although the final Kyoto target was not as strong as AOSIS would have liked, AOSIS believed that without their advocacy of a much stronger target, the Kyoto target would have been considerably weaker. Maldives President Gayoom noted 2 years later that the existence of the Kyoto Protocol itself is in part a result of AOSIS advocacy for an international treaty to address global warming. He commented, "I think our work has resulted in the Kyoto Protocol in which all states agreed to reduce their emissions of greenhouse gases to a certain extent—although not 20% by the year 2005. But this is still a good step forward."

AOSIS staked out a principled and strong position early in the process, and this widened the framework for debate. The environmental NGOs in the Climate Action Network endorsed this target and created political momentum in Japan, Europe and the United States for the AOSIS position, and this created valuable pressure on negotiators from these countries. Since a review of the adequacy of commitments was built into the Protocol, it will always be possible to strengthen the targets in the future if they are proven to be inadequate.

Although developing countries do not have legally binding emission-reduction targets in the Kyoto Protocol, Article 10 does call on all Parties to work on steps toward "cost-effective national, and where appropriate, regional programs to improve the quality of local emission factors." This too was consistent with AOSIS interests, because many small island states wanted to do what they could to reduce greenhouse gas emissions in their countries.

Initially the implementation mechanisms were met with reservations from members of AOSIS. "Nobody knew what trading meant," said Chairman Slade, "By their very description, nobody knew how (the mechanisms) would work and how they would control emissions." Although their qualms were not completely quelled by the end of the Kyoto summit, there would be an opportunity to shape these mechanisms in future negotiations, because the details had not been worked out in Kyoto. In addition, the compliance mechanism was left unfinished, but AOSIS had managed to make sure that such a mechanism would exist. In fact, Marshall Islands negotiator Espen Ronneberg became the Co-Chairman of the working group on the compliance mechanism in the first round of negotiations after Kyoto, after demonstrating knowledge and skill on the compliance issue during the Kyoto talks.

9.4 Negotiation Lessons Learned

- **Preparation**—AOSIS rigorously prepared for each negotiation session and developed briefing books for its members. They focused on helping each other find their strengths and in doing so capitalized on their resources. Before and during negotiations, they would often bring in experts on specific topics to brief them in detail.
- **Coordination**—By using a coordinated approach, AOSIS was able to participate in every important meeting during the negotiations, many of which occurred concurrently. "You can't afford not to have people at a meeting. You have to have people everywhere," explains Ambassador Slade.
- **The Power of a Good Proposal**—By putting the first detailed and careful proposal on the table, AOSIS shaped the form and content of the eventual Kyoto Protocol. Also, because AOSIS made an aggressive proposal regarding the emissions-reduction target, they widened the framework for debate and created a "pulling" force for stronger emissions reductions than probably would have occurred otherwise. The environmental NGO community publicly endorsed the AOSIS target and created a media campaign to motivate public pressure for this position.
- **Coalition-Building**—AOSIS was itself a coalition of small, relatively powerless countries. Individually, these countries would probably not have been listened to during the course of the actual negotiations. As a bloc, however, they were regularly expected to comment on each step of the negotiations and thereby became a potent political force in the negotiations. AOSIS also learned to form coalitions with other blocs of countries, such as the G-77, to increase their negotiating power when possible.
- **Moral suasion**—AOSIS was skillful at reminding all the negotiators what was at stake—the very existence of their countries. They became effective at bringing recalcitrant negotiators back to the table and also at bringing pressure to bear through the media.
- **Flexibility**—Because coalitions were so important, AOSIS tried to be as flexible as possible so that it could retain alliances with other countries or blocs.
- **Individual skill and leadership**—Over time, several prominent leaders emerged within AOSIS who eventually became leaders of the entire Kyoto Protocol process. Certain negotiators became chairmen of key working groups (for example on compliance), and the Chairman of AOSIS, Ambassador Slade, was widely respected by all of the other negotiators within and beyond the AOSIS coalition.
- **Internal capacity-building**—AOSIS created mechanisms for training each other about the details and processes of the negotiations. New negotiators had a ready support network and talented guides to introduce them to other negotiators and to the process. More experienced negotiators were able to trust each other to negotiate and provide accurate and useful information.

- **Collaborating with NGOs**—AOSIS solicited the advice and technical assistance of a number of NGOs, including FIELD and Ozone Action. This helped them obtain additional legal expertise and publicity.
- **Power**—Although one would assume that small island states lack power, AOSIS demonstrates that skillful negotiation and coalition-building can offset a low-power situation. In fact, AOSIS often acted as a political counterweight to the OPEC countries, blocking the OPEC coalition from completely obstructing the process. Both AOSIS and OPEC were parts of the G-77 even though they fundamentally had different interests.
- **Linkage**—AOSIS accepted that there would be trade-offs in the negotiations. They knew that they would never win approval of a 20% reduction in emissions, for example, but by figuring out what they could be flexible about and what had to be part of the final package, they were able to meet more of their interests.

9.5 Addendum

AOSIS was very skillful in its negotiations; however, the survival of AOSIS countries remains very much in question. While the newfound commitment of large polluters such as the United States gives reason for optimism, AOSIS fears the new commitments are not enough. At the Major Economies Forum (MEF)—which included the G8, China, India, Brazil and Indonesia—leaders "agreed to cap the rise in the Earth's average temperature to 2°C (3.6°F) above eighteenth century levels" (AFP 2009). AOSIS, on the other hand, wants that cap set at 1.5°C, saying that a 2 degree rise is too high for the survival of their islands (AFP 2009).

AOSIS has responded by emphasizing that the survival of their countries is at stake, attempting to use moral pressure to persuade countries to go along with their goals. They have also emphasized the need for new objective criteria, arguing that the IPCC's fourth assessment report is no longer the most up-to-date material, and therefore no longer the best objective standpoint. Another proposition on the table is expanding the Montreal Protocol on Substances that Deplete the Ozone Layer (Gronewold 2009), which "accelerated the phaseout schedules of the controlled substances [...including] CFCs, halons, carbon tetrachloride, methyl chloroform, HCFCs, HBFCs, and methyl bromide" (AFEAS). The idea is to expand the protocol so that it includes reducing hydrofluorocarbons as well (Gronewold 2009).

As more countries support AOSIS's general cause, it should be easier for AOSIS to persuade those countries to side with their specific positions. The key is for AOSIS to continue to create good relationships with countries and assert the common interests of these nations. And while the MEF may not have produced the results that AOSIS had wanted, the negotiations represent a step in the right direction. Moreover, in his address at the MEF, United States President Barack Obama said, "developed countries, like my own, have a historic responsibility to take the lead. We have the much larger carbon footprint per capita. And I know that in the past the United States has sometimes fallen short of meeting our responsibilities. So let me be clear: Those days are over" (Obama 2009). Not only does Obama's quote indicate that the United States, a country with a huge carbon footprint, will be reducing its emissions, it also means that Obama has a huge personal interest in limiting climate change and that he and other developed countries want to be global leaders on climate change. AOSIS can show these countries that for them to be leaders, they need to make even greater strides than at present, meanwhile catering to these countries' egos by stating trust that they will do what is necessary and they will help ensure the survival of these small island states.

AOSIS has garnered support from many nations. Norway, for example, pledged $260,000 to AOSIS for their climate change efforts, which allows AOSIS to establish a secretariat in New York (Norway Mission to the UN). Iceland has declared that it will reduce its own emissions (despite contributing less than 0.1% of global emissions) because it does not believe any nation should be exempt from the efforts. International support for climate change efforts are reason for optimism for AOSIS.

Still, with the clock ticking, there is no guarantee that these efforts will be enough for island states to survive, which means alternative plans are needed. The Republic of Kiribati, an AOSIS member, has one idea. In a speech at Harvard University (September 2008), Kiribati President Anote Tong said that he is looking for ways for his country's citizens to migrate to other countries, such as Australia, where they can work and live as "climate refugees." AOSIS is working to ensure that such measures may not be necessary, but such an idea demonstrates the laborious preparation of AOSIS, which has made great strides in the fight against climate change and the efforts to ensure the survival of their nations.

Chapter 10
Conclusion: Putting It All Together

Abstract "Achieving better health outcomes for all." This is a noble endeavor, and in an increasingly interdependent world this goal can no longer be accomplished by only those who work in public health nor through more traditional avenues of increasing access to primary health care or augmenting health budgets. To address the major health crises of today and to prevent or mitigate them in the future requires all of us to seek collective agreement and actions within and across countries. As we have demonstrated throughout this guide, effective collective action is achieved by developing and implementing comprehensive negotiation strategies. And these actions must be taken both within and outside of the health sector by a range of actors, not simply health policy-makers.

Keywords Health policy makers · Better health outcomes · Framework for negotiation · Issue framing · Joint fact-finding · Mutual gains approach · Coalition building · Implementation · Institutional change · Interpersonal relationships · Culture · Gender

Once thought to be the purview of only international diplomats, negotiation skills are now recognized as a critical part of any policy-makers' toolbox. The actors that may be involved in negotiations include diplomats, ministers, health policy-makers, foreign policy-makers, trade negotiators, health practitioners, and program managers within donor, governmental and nongovernmental organizations. Negotiations will take place in a dizzying number of different contexts, from the WHO to the WTO, from a meeting on the MDGs to a conference on climate change, from a team meeting within the Ministry of Health to a donor pledging conference.

D. Fairman et al., *Negotiating Public Health in a Globalized World*,   161
SpringerBriefs in Public Health, DOI: 10.1007/978-94-007-2780-9_10,
© The Author(s) 2012

Box 10.1 Though the FCTC resulted from a drive to protect health and to further healthy lifestyles, the work of the Intergovernmental Negotiating Body on the protocol involves many sectors other than health. The participants in the second session, for instance, included ambassadors and heads of mission along with delegates from ministries of foreign affairs, finance, trade, justice and health, plus representatives of customs and taxation departments. A number of delegations included lawyers, the European Union's three top delegates were from the European Anti-Fraud Office, and one national delegation was headed by a senior figure from the country's state tobacco monopoly.

In all these situations, negotiators will be called upon to manage all or some aspects of the negotiation process. And, at first glance, the process can seem complex. As we now understand, each of us has some goals and interests at stake in the negotiation. None of us can get what we want without some cooperation from the other. Each of us would like to gain as much as possible from the negotiation process, but we must balance this desire with the requirement to provide something of value to our negotiation partner, and with a realistic assessment of how our interests will be affected if we do not reach agreement.

Combined with the limited information we have about the others' true goals and interests, it is not always obvious what to offer, how to offer it or how to find out what would be worth offering. The way we communicate with each other can have a significant and often unintended impact on the outcome. And the relationships we form or develop during the negotiation process can have a significant impact not only on the present negotiation, but also on potential future negotiations with these parties and with others.

The most effective way to manage the complexity and the multiple relationships is to use a framework for negotiations. The five tasks highlighted in this guide constitute an iterative and reinforcing set of processes that can help negotiators to better manage these important negotiations. Here is a recap of those tasks, with key questions and actions for negotiators and their organizations:

1. Issue framing and forum choice

   - WHO—Are you targeting the right stakeholders for action (i.e., people who have influence or decision making authority over the issue)?
   - HOW—Is your message crafted for maximum influence (i.e. in terms that are compelling to the other stakeholders because they appeal to their interests, fears or moral values, and that provide solutions, not just problems)?
   - WHEN—Is the timing of your initiative optimized to build maximum momentum (through linkage with other global priorities and resources, taking advantage of windows of opportunity created by dramatic development)?
   - WHERE—Have you identified a forum for negotiation that is likely to favor your interests, whose decisions or agreements are likely to have a significant

impact on the issue, and where you can participate directly or have strong allies as your representatives?

2. Managing the negotiation process, including strategies for:

a. Joint fact-finding

- Have you worked with other stakeholders to agree on technical/scientific questions?
- Have you jointly identified and selected qualified resource people/experts?
- Have you jointly monitored the fact-finding process?
- Have you jointly reviewed and discussed the results of the process?
- Have you resolved issues of fact in a way that enables the process to focus primarily on interests and options to address a well-understood situation?

b. Employing a mutual gains approach to negotiation

- **Preparation:** Have you defined your own interests and BATNA; estimated other negotiators' interests and BATNAs; and prepared options to offer and questions to ask others?
- **Creating Value**: Have you moved beyond positions to explore interests and priorities, and invented mutually beneficial options without committing?
- **Reaching agreement**: Have you used a single-text approach to clarify areas of agreement and disagreement? Have you made trades across issues negotiators value differently? Have you evaluated options using "objective" criteria? Have you used contingent agreements to manage different views of the future? Have you used consensus and fallback voting rules to make final decisions?

3. Coalition-building

- Does joining a coalition increase your ability to influence the negotiation on your top priority interests?
- Does joining a coalition pose any risks, either of being forced to trade-off important interests for the sake of coalition unity, or of alienating potential allies who have negative relationships with the coalition?
- Map the negotiating environment comprehensively to determine potential coalition partners: who are the parties at and away from the table? What are their primary interests? How substantial is their influence on the outcome of the negotiation? Looking at your stakeholder map, who are the most likely coalition partners, and who might form a blocking coalition?
- How can you best reach out to engage with influential stakeholders whose interests are not entirely aligned with yours? What potential partners should you approach and in what sequence ($\Rightarrow$ backward mapping)?
- How can you deal with blocking coalitions: through mutual gains negotiation, constructive trade-offs, and/or strategic action to split the coalition?
- How can you create synergy and complementarity among government, NGO, business and multilateral actors involved in the negotiation process?

4. Managing implementation challenges

- Does this agreement satisfy all or nearly all of the key stakeholders' primary interests well enough so that they have incentives to comply without further enforcement mechanisms?
- Are there mutually acceptable ways to monitor implementation—both to verify that parties are meeting their commitments and to confirm that the actions they are taking are having the desired impact on the problem?
- Does the agreement include contingency provisions that anticipate and address potential implementation challenges or changes in circumstances that may affect the agreement?
- Does the agreement include procedures or mechanisms to assist the parties in answering questions and resolving disputes about implementation?
- Do the implementation stakeholders (agencies and organizations that will be directly responsible for taking action) have the resources, capacities and incentives necessary to take the actions that they would be committed to under this agreement?
- If not, does the agreement provide a mechanism to mobilize resources, build capacities and/or change incentives?

5. Institutional change for effective negotiation

- Have you identified influential sponsors and champions for building institutional negotiation capacity?
- Is there a shared model and language for negotiation for the organization?
- Has a critical mass of people been trained in mutual gains negotiation, tailored to the organization? Do they have opportunities, incentives and supports to put the training into action?
- Are there templates and tools to help staff at all levels apply negotiation concepts and skills effectively?
- Have organizational incentives, procedures and resources been reviewed and modified to measure, support and reward effective negotiation?
- Are there clear metrics for negotiation success, and processes for 360 evaluation and feedback to improve negotiator performance?

The authors believe that the framework presented in this book will provide a strong foundation for governmental, civil society, business and multilateral agency stakeholders who want to improve their outcomes in global health negotiations. At the same time, we acknowledge that other topics are important to the negotiation process and are worthy of further study by negotiators and advocates. Two topics are of particular importance to consider further:

Interpersonal relationship building: We have touched very lightly on the importance of building relationships based on trust, and on addressing past problems in relationships in order to build effective partnerships for the future. There is a useful literature on relationship building in the context of negotiation, including William Ury, *Getting Past No* (New York: Bantam 1991), Doug Stone,

Bruce Patton and Sheila Heen *Difficult Conversations* (New York: Penguin 1999), and Roger Fisher and Daniel Shapiro, *Beyond Reason* (New York: Penguin 2005). There is also a more general literature on emotional intelligence, including Daniel Goleman's seminal *Emotional Intelligence* (New York: Bantam 1995). We recommend a review of this literature for all negotiators who must manage the interpersonal dimension along with organizational, sectoral, national and global dimensions.

Culture and gender: There is significant evidence that a person's local national, organizational and disciplinary background, and the cultural norms of particular places, countries, organizations and disciplines, shape that person's approach to the negotiation process. Culture by definition is a set of shared meanings and assumptions within a group, influencing interpretations of spoken language (with the added challenge of translation in multi-lingual negotiations), body language, time, relationship, value and commitment, among other factors (*see* Cohen 1994, Chap. 2). Given the enormous diversity of cultural interpretations of these factors, the main advice that we and others in our field give to those involved in cross cultural negotiations is to take extra time to understand each other's core assumptions; test communication and understanding rather than assuming that the meaning behind spoken words will be interpreted as intended; and be sensitive to cultural differences in the way negotiators express interests and concerns, ask questions about others' interests, propose and test options, use third parties, confirm and ratify agreement, plan for and carry out implementation.

Likewise, there is evidence that gender and gender relations, defined differently in each of these cultural contexts, can affect negotiations (see for example Kolb and Williams 2000). However, we see culture and gender as highly context-specific variables in negotiation, and do not believe that there is very useful prescriptive advice about "how to negotiate with people from country X," "how to negotiate with women," or "how to negotiate with men."

Promoting health for a country's population is not simply within the hands of that government alone. As we have written elsewhere: "The increase in cross-border health risks has been accompanied by a decrease in national governments' capacity to respond to them effectively.... Health problems and the keys to their resolution now cut across national boundaries and often need international global solutions" (Chigas et al. 2007). The scope and depth of these negotiations are daunting, but the challenges in managing them are not insurmountable. Articulating priority issues, understanding how the negotiation process works, building coalitions within and across organizations and even national borders, and developing the architecture for effective implementation are all critical—and manageable—pieces. This guide has offered a framework for navigating the global health diplomacy process, and we encourage readers to turn immediately to the toolbox contained in the appendices to enhance their mastery of the skills.

Appendix 1: Key Concepts and Frameworks in Global Public Health Negotiations

Joint Fact-Finding: A process in which stakeholders work jointly to define the technical and/or scientific questions to be answered and to identify and select qualified resource persons from epistemic communities to assist the group. In collaboration with these resource persons, the stakeholders then refine the questions; set the terms of reference for technical or scientific studies; monitor, and possibly participate in, the study process; and review and interpret the results. Thus, they work together to establish a common set of facts.

Consensus-Building: A framework for decision-making in which groups seek representation of all affected stakeholders; gain a shared understanding of each others underlying interests and of the technical, political, social, economic and environmental issues at stake; jointly develop options that are more creative and widely supported than the initial proposals of any one stakeholder; and seek agreements that satisfy everyone's primary interests. A consensus-building process includes six distinct stages: assessing the potential; designing and deciding on a process; clarifying facts and options; seeking joint gains; reaching agreement; and implementing, adapting and learning from the process.

Mutual Gains Approach: A four-step approach to negotiation that greatly improves negotiation capacity through (1) effective preparation; (2) value creation, by focusing on needs and interests rather than positions, inventing options and proposals that meet all stakeholders' needs and interests and thus seeking to maximize joint gains before deciding "who gets what;" (3) value distribution, using mutually acceptable criteria; and (4) effective implementation, by ensuring the sustainability of agreements through commitments to continue communication, joint monitoring, contingency planning and dispute resolution mechanisms.

Positions and Interests: A *position* is the stance one takes on an issue (e.g., "we are going to allow domestic drug producers to manufacture generic drugs without first obtaining licenses"). In contrast, an *interest* is a core need or want that underlies a position (e.g., "we want access to affordable, life-saving drugs for large and impoverished parts of our population"). Interests can concern the substance of

D. Fairman et al., *Negotiating Public Health in a Globalized World*,
SpringerBriefs in Public Health, DOI: 10.1007/978-94-007-2780-9,
© The Author(s) 2012

an issue; the timing of action to be taken; the level of risk that can be assumed; the level of consideration or setting of precedents; reputational issues; or simply the way initiatives may be framed.

Best Alternative to a Negotiated Agreement (BATNA): The best solution *away* from the negotiating table in the absence of reaching agreement. The BATNA is the best of the range of fallback options in the event that the negotiation does not produce a package that meets one's most important interests and one's bottom line for the negotiation.

Zone of Possible Agreement (ZOPA): The range of negotiation outcomes that are acceptable to all stakeholders who need to consent.

Coalition: A group of parties that collaborate to advance shared interests.

Issue Mapping: A tool for analyzing the issues at stake in a negotiation and the respective probable and actual positions, interests and priorities of the different stakeholders.

Option Mapping: A tool for analyzing the different stakeholders' BATNAs to identify the ZOPA and options for agreement.

Backward Mapping: A tool to assist in the process of identifying and targeting stakeholders for action. Starting with the desired outcome, the negotiator identifies key players, guided by the question: "Which parties are needed or at least helpful to achieve each preceding step that is required to move toward the outcome?" One starts by asking, "What decision or action do we seek?" and "Who needs to take those decisions or actions." Then, "Can we reach and influence those actors directly?" If so, one can begin developing a negotiation strategy. If not, ask, "Who could directly communicate with and influence those decision-makers and actors?" and ask again "Can we reach and influence those actors directly?" The negotiator must keep going until a pathway has been mapped from those who can be influenced directly to those who are the ultimate decision-makers.

Appendix 2: Tools for Preparing and Conducting Negotiations

1. **Preparation Tool** for Mutual Gains Negotiations
2. **Stakeholder Mapping**: Mapping Key Stakeholders to Identify Potential Coalitions
3. Overcoming Stalemate: **Currently Perceived Choice Tool**

D. Fairman et al., *Negotiating Public Health in a Globalized World*,
SpringerBriefs in Public Health, DOI: 10.1007/978-94-007-2780-9,
© The Author(s) 2012

Tool A.2.1: Negotiation Preparation Worksheet

INTERESTS		ALTERNATIVES		OPTIONS
Our Interests	Their Interests	Our BATNA	Their BATNA	Options for mutual gain
What are our interests in this negotiation? What are our hopes, concerns, fears, desires and wants on the issues being negotiated?	*What are their interests in this negotiation? What are their hopes, concerns, fears, desires and wants on the issues being negotiated?*	*What is our Best Alternative to a Negotiated Agreement—BATNA?* *What will we do to meet our interests in the absence of an agreement?* *How might we improve our BATNA?*	*What is their Best Alternative to a Negotiated Agreement—BATNA?* *What will they do to meet their interests in the absence of an agreement?* *How might we weaken their BATNA or raise doubts about how realistic they are being about their BATNA?*	*What options might meet their interests well and our interests very well, and be better than both BATNAs?* *Have we considered all sources of value creation (e.g., differences in priorities, differences in capabilities, differences in assessments of risk, etc.)?*

What interests are:
- *Shared?*
- *Opposed?*
- *Just different (but reconcilable)*

(continued)

Tool A.2.1: (continued)

OBJECTIVE CRITERIA	COMMITMENT	SUSTAINABILITY AND IMPLEMENTATION
What arguments/criteria/reasons can we give for preferring the option (package) that is best for us?	*What authority do I/we (negotiators) have to make firm commitments in the upcoming negotiation? What mandate/authority do I need in the upcoming meeting to be effective?*	*What implementation problems are likely to arise if they accept our proposal/preferred option?*
How can we help them "sell" this option to their bosses or constituents?	*Who else needs to approve or support the negotiation analysis above?*	*How might implementation problems be avoided or overcome?*
What criteria will seem fair to them?		*- Can we make the agreement "nearly self-enforcing" (i.e. provide incentives for parties to live up to their commitments)?*
		- How will we monitor implementation? Resolve disputes over interpretation of monitoring data?
		- Deadlines or milestones for reaffirming or reconsidering commitments?
		- What dispute resolution mechanisms will we put into place to deal with implementation problems?
		- How can we build trust and establish good communication to make it easier to deal with problems that arise?

STAKEHOLDER MAPPING FOR COALITION IDENTIFICATION

Fig. A.2.1 Stakeholder mapping for coalition identification

Tool A.2.2: *Mapping Key Stakeholders to Identify Potential Coalitions*

The following "stakeholder mapping tool" provides a systematic way of mapping key parties, issues and interests to aid negotiators in deciding whether or not to join or form a coalition. Using this tool, the negotiator can do a quick initial assessment of the interests and influence of other stakeholders with regard to the goals that the negotiator is trying to achieve. This will help to identify potential coalition partners who share interests, blocking coalitions that might need to be addressed, and other stakeholders who might be mobilized in support or opposition.

(1) Identify the major stakeholders in the negotiation process. Include not only the parties at the negotiating table, but major factions or domestic stakeholders within a party who may have a significant influence on the negotiation, as well as non-governmental parties such as NGOs or corporations who are influential.

(2) Map the stakeholders in the quadrants of the tool, according to their level of influence on the negotiation and their level of support for or convergence with your interests. Stakeholders who can influence the negotiations significantly and who share your interests (upper right hand quadrant) will likely be your top priority allies and coalition partners. Those who may have significant influence on the negotiation but oppose your interests, or may not support them because the issues in question are not of high priority for them (upper left hand

quadrant) are potentially members of a blocking coalition, unless some combination of education, advocacy and/or negotiation can shift them to support the coalition. Stakeholders who do not share your interests but also have little influence on the negotiation (lower left hand quadrant) may be mobilized to support a blocking coalition, but may also be persuaded to support the negotiator's goals, or may be ignored. Finally, those who support your interests but have little influence on the outcome of the negotiation (lower right hand quadrant) stakeholders may be mobilized to support your coalition.

(3) Once you have completed your stakeholder map, you can begin to plan a sequence of meetings to either build support for your issue(s) or mitigate opposition to your position, depending on your goals. Keep in mind that coalitions can shift over time, as issues evolve and circumstances change, so the stakeholder map should be kept up to date throughout the negotiation process.

Tool A.2.3: Currently Perceived Choice Tool (CPC Tool)

The purpose of this tool is to give us a clear and empathetic understanding of why someone is now saying "no" when we want them to say "yes." We need to know where their mind is today if we hope to change it tomorrow. This tool helps understand the target decision maker(s)' underlying motivations, perceptions and choices so that one can frame the problem in a way that is persuasive *to them*. It asks four questions:

(1) Who is the target decision maker? Who needs to take the decisions or actions we seek? Decisions are made by individuals (or groups of individuals), not organizations. And even if some group of people must reach a consensus on a decision, someone must put that decision before the group.

(2) What is the question *that decision maker(s)* see themselves being asked? The challenge here is to capture the question the decision maker perceives, not the question we wish they would perceive.

(3) What are the consequences *for the decision maker(s)* of saying "yes" to the decision or action they perceive they are being asked to take? What kinds of things would they fear or imagine might happen—to them, to their constituents, to their organization, etc.—if they were to say "yes?" If the decision makers are already not taking or refusing to take the desired action, one can assume that they perceive the *negative* consequences of saying "yes" as outweighing the positive.

(4) What are the consequences, from the decision maker(s)' perspective, of saying "no?" Again, if they are refusing to take action or accept a proposal, it is likely that they perceive positive consequences of saying "no," or, at a minimum, that they can wait to do something later.

If, after reflecting on their CPC, we can understand how it makes sense for the other party(ies) to say "no" to their proposal. The next step is to design a new choice for them, one to which the answer "yes" is more likely—a target future choice. Use the same tool format to identify the characteristics of what such a new choice might be.

(1) Begin with a generic new question. Because it is not clear yet what the new question or proposal should be, it is useful to focus first on the perceptions that it should produce in the decision maker's head. The new question might be as general as, "Should I accept the 'X Plan?'"

(2) Create "yes" and "no" columns, just as in the Currently Perceived Choice. Here, however, imagine what interests of the decision maker might need to be met in order for the decision maker to say "yes." What positive consequences of saying "yes" would be persuasive to the decision maker? And how would the decision maker need to perceive the consequences of saying "no" in order to be persuade that rejection is not good for him or her? Use the Currently Perceived Choice as a guide: how might concerns that currently lead them to say "no" be handled differently in the future to produce a "yes?"

(3) Brainstorm possible options that could be "yesable" to the decision maker, i.e., that might be perceived in the way you have outlined in the target future choice?

CURRENTLY PERCEIVED CHOICE	
Decision maker: _____	
Question: _____	
If "Yes" *Perceived consequences to decision maker of saying "yes"*	If "No" *Perceived consequences to decision maker of saying "no" (his/her BATNA)*
-	+
-	+
-	+
-	+
BUT (*some positive consequences of saying "yes," but as the decision maker is rejecting our proposal, they likely do not outweigh the negative consequences*)	BUT (*some negative consequences of saying no, but as the decision maker is already saying "no," they are likely not as important as the positive consequences*)
+	-
+	-
NEVERTHELESS:	NEVERTHELESS:
-	+
-	+

References

Annan, K. (2001). Speech proposing global fund for fight against HIV/AIDS and other infections diseases. AFR/313/Rev.1, printed in United Nations Press Release SG/SM/7779/Rev.1. (http://www.un.org/News/Press/docs/SGSM7779R1.doc.html).

British Broadcasting Company. (2005). Brazil turns down US Aids funds. http://news.bbc.co.uk/2/hi/americas/4513805.stm. Accessed 21 July 2009.

BBC News Online. (2010). CRU climate scientists 'did not withhold data.' http://www.bbc.co.uk/news/10538198. Accessed 24 May 2010.

BBC News Online. (2006). South Africa AIDS policy attacked. http://news.bbc.co.uk/2/hi/africa/5265432.stm. Accessed 20 Nov 2010.

BBC News Online. (2006). U.S. criticized for HIV aid effort. http://news.bbc.co.uk/2/hi/health/4797537.stm. Accessed 20 Nov 2010.

Bill & Melinda Gates Foundation. (2002). *Developing successful global health alliances*. Seattle: Bill & Melinda Gates Foundation. http://www.gatesfoundation.org/global-health/Documents/GlobalHealthAlliances.pdf. Accessed 24 May 2010.

Blouin, C. (2007). Trade policy and health: From conflicting interests to policy coherence. *Bulletin of the World Health Organization, 85*(3), 170.

Chigas, D., et al. (2007). Negotiating across boundaries: Promoting health in a globalized world. In C. Blouin, J. Heymann, & N. Drager (Eds.), *Trade and health: Seeking common ground*. Montreal: McGill-Queen's University Press.

Chigas, D. (1996). Preventive diplomacy and the OSCE: Creating incentives for dialogue and cooperation. In A. Chayes & A. H. Chayes (Eds.), *Preventing conflict in the post-communist world: Mobilizing international and regional organizations*. Washington DC: Brookings Institution.

Clinton, W. (2000) Letter to senator Dianne Feinstein on signing an executive order on access to HIV/AIDS pharmaceuticals and medical technologies. *Weekly Compilation of Presidential Documents, 36*(19), 1058. (http://frwebgate4.access.gpo.gov/cgibin/PDFgate.cgi?WAISdocID=218981657+0+2+0&WAISaction=retrieve\).

Cohen, R. (1994). *Negotiating across cultures: International communication in an interdependent world*. Washington D.C.: U.S. Institute of Peace.

Collin, J. (2004). Tobacco politics. *Development, 47*(2), 93.

Collin, J., Lee, K., & Bissell, K. (2002). The framework convention on tobacco control: The politics of global health governance. *Third World Quarterly, 23*(2), 275.

Cooper, R. (2010) Europe's Emissions Trading System. U.S. Climate Task Force, June 2010. (http://www.climatetaskforce.org/2010/06/22/europes-emissions-trading-system/).

Correa, C. M. (2006). Implications of bilateral free trade agreements on access to medicines. *Bulletin of the World Health Organization, 84*(5), 399.

Dodgson, R., Lee, K. & Drager, N. (2002). *Global health governance: A conceptual review.* Geneva: World Health Organization and London School of Hygiene and Tropical Medicine. http://cgch.lshtm.ac.uk/globalhealthgovernance.pdf. Accessed 19 April 2010.

Drager, N., & Fidler, D. P. (2007). Foreign policy, trade and health: At the cutting edge of global health diplomacy. *Bulletin of the World Health Organization, 85*(3), 162.

Drager, N., & Sunderland, L. (2007). Public health in a globalising world: The perspective from the world health organization. In A. F. Cooper, J. J. Kirton, & T. Schrecker (Eds.), *Governing global health: Challenge, response, innovation* (pp. 67–78). Hampshire, UK: Ashgate Publishing Ltd.

Ehrmann, J. R., & Stinson, B. L. (1999). Joint fact-finding and the use of technical experts. In L. Susskind, S. McKearnan, & J. Thomas-Larmer (Eds.), *The consensus building handbook: A comprehensive guide to reaching agreement.* Thousand Oaks, CA: Sage Publications.

Eilperin, J. (2007). U.S., China got climate change warnings toned down. *Washington Post.* (http://www.washingtonpost.com/wp-dyn/content/article/2007/04/06/AR2007040600291.html).

Fidler, D. (2007). Achieving coherence in anarchy: Foreign policy, trade and health. In C. Blouin, J. Heymann, & N. Drager (Eds.), *Trade and health: Seeking common ground.* Montreal: McGill-Queen's University Press.

Fisher, R., Ury, W., & Patton, B. (1991). *Getting to yes: Negotiating without giving in* (2nd ed.). Cambridge, MA: Houghton Mifflin.

Foege, W. H. (1998). Global public health: Targeting inequities. *Journal of the American Medical Association, 279*(24), 1931–1932.

Gallagher, K. S., & Berland, A. (2001). *Keeping your head above water in climate change negotiations: Lessons from island nations.* Cambridge, MA: Unpublished manuscript. Available from Consensus Building Institute.

Garrett, L. (2007). The challenge of global health. *Foreign Affairs, 86*(1), 14–38.

Global Health Council. (2005). *Strengthening the message: Building support for maternal and child health.* http://www.globalhealth.org/images/pdf/communication_tools/strengthening_message.pdf. Accessed 20 Nov 2010.

Goodin, R., & Dryzek, J. (2006). Deliberative impacts: The macro-political uptake of mini-publics. *Politics and Society, 34*(2), 219–244.

Hearns, W., Nixon, D. (2007). Rotary international, gates foundation commit $200 million. *Rotary International News,* . http://www.rotary.org/en/MediaAndNews/News/Pages/071120_news_gannounce.aspx. Accessed 20 Nov 2010.

Kickbusch, I., Silberschmidt, G., & Buss, P. (2007). Global health diplomacy: The need for new perspectives, strategic approaches and skills in global health. *Bulletin of the World Health Organization, 85*(3), 231.

Kolb, Deborah, & Williams, J. (2000). *The shadow negotiation: How women can master the hidden agendas that determine bargaining success.* New York: Simon & Schuster.

Kotter, J. (1996). *Leading change.* Boston: Harvard Business School Press.

Lakshmanan, Indira, A. R. (1997, December 11). Accord set on cutting emissions; 160 nations in agreement; US hails 'historic first step'. *The Boston Globe,* A1.

Lax, D., & Sebenius, J. (2006). *3D negotiation: Powerful tools to change the game in your most important deals.* Boston: Harvard Business School Press.

Lewis, N. (2000). Clinton issues order to ease availability of AIDS drugs in Africa. *New York Times.* http://www.nytimes.com/2000/05/11/world/clinton-issues-order-to-ease-availability-of-aids-drugs-in-africa.html?pagewanted=1. Accessed 20 Nov 2010.

Martinez, J. (2001). *Negotiations: Negotiating the cartagena biosafety protocol to the convention on biological diversity.* Cambridge, MA: Unpublished manuscript. Available from Consensus Building Institute.

Martinez, J., & Susskind, L. (2000). Parallel informal negotiation: An alternative to second track diplomacy. *International Negotiation*, *5*(3), 569–586.

McKercher, B., & Chon, K. (2004). The over-reaction to SARS and the collapse of Asian tourism. *Annals of Tourism Research*, *31*(3), 716–719.

Médecins Sans Frontières. (2001) *US action at WTO threatens Brazil's successful AIDS program.* New York/Geneva: Press Release.

Ministers of Foreign Affairs of Brazil, France, Indonesia, Norway, Senegal, South Africa and Thailand. (2007). Oslo ministerial declaration: Global health—a pressing foreign policy issue of our time. *The Lancet 369*(9570): 1373-1378. http://dx.doi.org with DOI 10.1016/ S0140-6736(07)60498-X

Minzter, I., & Leonard, J. A. (Eds.). (1994). *NNegotiating climate change: The inside story of the rio convention.* Cambridge: Cambridge University Press.

Mnookin, Robert H. (2003). Strategic barriers to dispute resolution: A comparison of bilateral and multilateral negotiations. *Journal of Institutional and Theoretical Economics*, *159*, 199.

Movius, H., & Susskind, L. (2009). *Built to win: Creating a world class negotiating organization.* Boston: Harvard Business School Press.

Musungu, S.F., Oh, C. (2005). *The use of flexibilities in TRIPS by developing countries: Can they promote access to medicines? Study 4C.* Geneva: Commission on Intellectual Property Rights, Innovation and Public Health, World Health Organization. http://www.who.int/ intellectualproperty/studies/TRIPS_flexibilities/en/index.html.

Ng, S. (2003). Chinese tourism recovering from SARS. *Asia Times Online* (http://www. atimes.com/atimes/China/EG11Ad02.html).

Nunn, A. (2009). *The politics and history of AIDS treatment in Brazil.* New York: Springer Science and Business Media, LLC.

Organisation for Economic Co-operation, Development (OECD). (2005). *BBridge over troubled waters: Linking climate change and development.* Paris: OECD.

Owen, J.W., Roberts, O. (2005) Globalization, health and foreign policy: Emerging linkages and interests. *Globalisation and Health* 1, 12. http://www.globalizationandhealth.com/content/ 1/1/12.

Oxfam, G. B. (2001). *WTO patent rules and access to medicines: The pressure mounts. Policy briefing.* Oxfam GB: Oxford.

Petersen, M., Rohter, L. (2001). Maker agrees to cut prices of 2 AIDS drugs in Brazil. *New York Times*, 2001. http://www.nytimes.com/2001/03/31/health/31AIDS.html?pagewanted=print.

Q&A: SARS. (2004). *BBC News Online.* http://news.bbc.co.uk/2/hi/health/2856735.stm.

Reel, M. (2006). Where prostitutes also Fight AIDS: Brazil's sex workers hand out condoms, crossing U.S. ideological line. *Washington Post.* Retrieved From http://www.washingtonpost. com/wp-dyn/content/article/2006/03/01/AR2006030102316_pf.html.

Reinhardt, E. (2006). Access to medicines. *UN Chronicle Online Edition*, 43(3). http://www.un.org/Pubs/chronicle/2006/issue3/0306p56.htm

Ribando, C. (2007). Brazil-U.S. relations. *Congressional Research Service.* Retrieved From http://www.wilsoncenter.org/news/docs/RL33456.pdf.

Roemer, R., Taylor, A., & Lariviere, J. (2005). Origins of the WHO framework convention on tobacco control. *American Journal of Public Health*, *95*(6), 936–938.

Sachs, J. (2001). *Macroeconomics and health: Investing in health for economic development. Report of the commission on macroeconomics and health.* Geneva: World Health Organization.

Sebenius, J. K. (1984). *Negotiating the law of the sea: Lessons in the art and science of reaching agreement. Harvard economic studies series.* Boston: Harvard University Press.

Shadlen, K. C. (2004). Patents and pills, power and procedure: The North-South politics of public health in the WTO. *Studies in Comparative International Development*, *39*(3), 79–80.

Shiffman, J. (2009). A social explanation for the rise and fall of global health issues. *Bulletin of the World Health Organization*, *87*, 608–613.

Shood, M. (2009). Lula, Obama quickly become fast friends. *Top News*. http://topnews.us/content/24726-lula-obama-quickly-become-fast-friends.

Stedman, S. (1997). Spoiler problems in peace processes. *International Security, 22*(2), 5–53.

United Nations. (1993). *Convention on Biological Diversity*. 1760 United Nations Treaty Series 143.

United Nations. (1992). *Framework Convention on Climate Change*. 1771 United Nations Treaty Series 107. U.N. Doc. A/AC.237/18 (Part II)/Add.1.

United Nations General Assembly. (2009a). 63rd Session. 2001–2010: Decade to Roll Back Malaria in Developing Countries, Particularly in Africa. A/RES/63/234 (13 March 2009). http://www.un.org/en/ga/63/resolutions.shtml

United Nations General Assembly. (2009b). 63rd Session. Prevention and control of non-communicable diseases. A/RES/63/33 (20 May 2010). http://www.un.org/en/ga/63/resolutions.shtml

United Nations General Assembly. (2007). 61st Session. 2001–2010: Decade to Roll Back Malaria in Developing Countries, Particularly in Africa. A/RES/61/228 (5 Feb 2007). http://www.un.org/Depts/dhl/resguide/r61.htm.

United Nations General Assembly. (2001a). 55th Session. 2001–2010: Decade to Roll Back Malaria in Developing Countries, Particularly in Africa. A/RES/55/284 (7 Sept 2001). http://www.un.org/depts/dhl/resguide/r55.htm.

United Nations General Assembly. (2001b). 55th Session. Declaration of Commitment on HIV/AIDS. A/RES/S-26/2 (2 Aug 2001). http://www.un.org/depts/dhl/resguide/r55.htm.

Wilson, D. (2005). New blood-pressure guidelines pay off—for drug companies. *Seattle Times*. http://archives.seattletimes.nwsource.com/cgi-bin/texis.cgi/web/vortex/display?slug=sick26m&date=20050626. Accessed 20 Nov 2010.

Wogart, J., Calcagnotto, G., Hein, W., von Soest, C. (2008). AIDS, access to medicines, and the different roles of the Brazilian and South African governments in global health governance. Working Paper 86. Hamburg: German Institute of Global and Area Studies.

World Health Organization (WHO). (2010) Expert Working Group on R&D Financing. http://www.who.int/phi/ewg/en/index.html. Accessed 2 Dec 2010.

World Health Organization (WHO). (2009). Global Strategy and Plan of Action on Public Health Innovation and Intellectual Property. WHA 62.16. (22 May 2009). 62nd World Health Assembly, Resolutions and Decisions, WHA 62/2009/REC/1, p. 29. http://apps.who.int/gb/or/e/e_wha62r1.html.

World Health Organization (WHO). (2008). 61st World Health Assembly. Global Strategy and Plan of Action on Public Health Innovation and Intellectual Property. WHA 61.21 (24 May 2008). Sixty-first World Health Assembly, Resolutions and Decisions. WHA61/2008/REC/1, p. 31. Geneva: WHO. http://apps.who.int/gb/or/e/e_wha61r1.html.

World Health Organization (WHO). (2007a). Pandemic influenza preparedness: sharing of influenza viruses and access to vaccines and other benefits. WHA 60.28 (23 May 2007), A60/VR/11. Sixtieth World Health Assembly, Resolutions and Decisions. WHASS1/2006 WHA60/2007/REC/1, p. 102. Geneva: WHO. http://apps.who.int/gb/ebwha/pdf_files/WHA60/A60_R28-en.pdf.

World Health Organization (WHO). (2007b). Jakarta Declaration on Responsible Practices for Sharing Avian Influenza Viruses and Resulting Benefits. http://ecdc.europa.eu/pdf/Jakarta_Declaration_March_28th_Annex_4.pdf.

World Health Organization (WHO). (2007c). Press Briefing. http://www.who.int/mediacentre/news/releases/2007/pr09/en/index.html.2006.

World Health Organization (WHO). (2006a, May 22–27). World Health Assembly Resolution 59.26 (International trade and health). Resolutions and Decisions, Fifty-ninth World Health Assembly. WHA59/2006/REC/1: 37. Geneva: WHO. Available at http://apps.who.int/gb/ebwha/pdf_files/WHA59-REC1/e/WHA59_2006_REC1-en.pdf. Accessed 20 Nov 2010.

World Health Organization (WHO). (2006b) Public health, innovation, essential health research and intellectual property rights: towards a global strategy and plan of action. WHA 59–24. Fifty-ninth World Health Assembly: Resolutions and Decisions. 34. http://apps.who.int/gb/ebwha/pdf_files/WHA59-REC1/e/WHA59_2006_REC1-en.pdf. Accessed June 27, 2010.

World Health Organization (WHO). (2005) International Health Regulations. Fifty-eighth World Health Assembly. WHA 58.3. http://www.who.int/ihr/9789241596664/en/index.html. Accessed 20 Nov 2010.

World Health Organization (WHO). (2003). *WHO Global Conference on Severe Acute Respiratory Syndrome (SARS): Where do we go from here? (Summary Report).* Geneva: WHO. http://www.who.int/csr/sars/conference/june_2003/materials/report/en/index.html. Accessed 20 Nov 2010.

World Health Organization (WHO). (2003b). *Framework convention on tobacco control. WHA 56.1 (21 May 2003). Fifty-Sixth world health assembly.* Geneva: WHO.

World Health Organization (WHO). (2001). Secretariat update on the WHO consultation on potential liability and compensation provisions for the framework convention on tobacco control. A/FCTC/INB2/5 Rev.1 (6 July 2001). http://apps.who.int/gb/fctc/PDF/inb2/einb25r1. pdf. Accessed 20 Nov 2010.

World Health Organization (WHO). (1999). Avoiding the Tobacco Epidemic in Women & Youth, *International Conference on Tobacco and Health*, Kobe: WHO/NCD/TFI/KOBE/99.4.

British Broadcasting Company. (2005). Brazil turns down US Aids funds. http://news.bbc.co.uk/2/hi/americas/4513805.stm.

Reel, M. (2006). Where Prostitutes Also Fight AIDS: Brazil's Sex Workers Hand Out Condoms, Crossing U.S. Ideological Line. *Washington Post.* Retrieved From http://www.washingtonpost.com/wp-dyn/content/article/2006/03/01/AR2006030102316_pf.html.

Babbitt, E.F., (1999). Challenges for international diplomatic agents. In: L. Mnookin., & R. Susskind. (Eds.), *Negotiating on behalf of others* pp. 135–150.

Cohen, R., (1997). *Negotiating across cultures* (2nd ed.). Washington, DC: U.S. Institute of Peace, 9 43, 215–226.

Druckman, D., Negotiating in the international context. In: I.W. Zartman., & J.H. Rasmussen (Eds.),*Peacemaking in international conflict.* Washington, D.C: U.S. Institute of Peace.

Fisher, R. Beyond YES. In: J.W. Breslin., & J.Z. Rubin (Eds.), *Negotiation Theory and Practice.* Cambridge: PON Books. 123–126.

Fisher, R., Ury, W.L., & Patton, B.M. (1991) *Getting to YES: Negotiating agreement without giving.* (2nd ed.). New York: Penguin Books.

Fisher, R., (1991) Negotiating power: Getting and using influence. In: J.W. Breslin & J.Z. Rubin (Eds.), *Negotiation theory and practice.* Cambridge: PON Books.

Fisher, R., (1991) Negotiating power: Getting and using influence. In: J.W. Breslin & J.Z. Rubin (Eds.), *Negotiation theory and practice.* Cambridge: PON Books. pp. 127–140.

Hampden-Turner, C., Trompenaars, A. (1991). *The seven cultures of capitalism.*New York: Doubleday. pp. 1–102.

Kelman, H.C. , (1993). Coalitions across conflict lines: The interplay of conflicts within and between the israeli and palestinian communities. In: J. Simpson & S. Worchel (Eds.), *Conflict between people and groups.* Chicago: Nelson-Hall. pp. 236–258

McCarthy, W. (1991). The role of power and principle in getting to YES. In: J.W. Breslin & J.Z. Rubin (Eds.), *Negotiation theory and practice.* Cambridge: PON Books. pp. 115–122.

Rubin, J.Z., & Sander, F.E.A. (1991). When should we use agents? direct v. representative negotiation. In: J.W. Breslin & J.Z. Rubin (Eds.), *Negotiation theory and practice.* Cambridge: PON Books. pp. 81–88.

Salacuse, J. W. (1997). After the contract, What? negotiating to work successfully with aforeign partner. *Canadian International Lawyer*, 2(4), 195–200.

Salacuse, J.W. (1999). How should the Lamb negotiate with the lion. In: D. Kolb (ed.), *Negotiation Eclectics* pp. 87–99.

Salacuse, J.W., (1999) Law and power in agency relationships. In: L. Mnookin., & R. Susskind (Eds.), *Negotiating on Behalf of Others* pp. 157–175.

Index